INTERIOR VIEW OF GRIST MILL.

THE

PRACTICAL AMERICAN

MILLWRIGHT AND MILLER:

COMPRISING

THE ELEMENTARY PRINCIPLES OF MECHANICS MECHANISM, AND MOTIVE POWER,

HYDRAULICS, AND HYDRAULIC MOTORS, MILL DAMS, SAW-MILLS, GRIST-MILLS, THE OAT-MEAL MILL, THE BARLEY MILL, WOOL CARDING AND CLOTH FULLING AND DRESSING, WINDMILLS, STEAM POWER, ETC.

BY

DAVID CRAIK,

MILLWRIGHT.

ILLUSTRATED BY NUMEROUS WOOD ENGRAVINGS AND FOLDING PLATES.

PHILADELPHIA:

HENRY CAREY BAIRD,

INDUSTRIAL PUBLISHER,

406 Walnut Street.

1870.

PHILADELPHIA:
COLLINS, PRINTER, 705 JAYNE STREET.

CONTENTS.

CHAPTER I.

MECHANICAL POWERS.

CHAPTER II.

FLY AND BALANCE WHEELS.

CHAPTER III.

TRANSMISSION AND TRANSPORTATION OF MOTIVE POWER.

CHAPTER IV.

PECULIARITIES AND PROPERTIES OF WATER.

CHAPTER XIV.

SAW-MILLS (*continued*).

CHAPTER XV.

THE CIRCULAR SAW-MILL.

CHAPTER XVI.

GRIST-MILLS.

CHAPTER XVII.

THE OATMEAL MILL.

CHAPTER XVIII.

THE BARLEY MILL.

CHAPTER XIX.

WOOL CARDING AND CLOTH FULLING AND DRESSING.

CHAPTER XX.

WINDMILLS.

CHAPTER XXI.

STEAM POWER.

APPENDIX.

MERCHANT BOLT.

INTRODUCTION.

THE occupation of a millwright differs from that of almost all other tradesmen and mechanics, in that he is compelled to accommodate his work to a greater variety of circumstances, conditions, and contingencies. For instance, a millwright is employed to build two saw-mills, one is to be in or near a city, convenient to foundries and machine-shops, where a great variety of water-wheels, and all the other machinery required for its construction can be obtained cheaply and in abundance. But the water power and site are valuable here, and he must select the water wheels, and adapt every part to do the required amount of work with the least possible waste of water and room, and all the machinery and many of the fixtures will be of iron. The other saw-mill is to be built back in the interior of the wilderness where all supplies must be packed and carried at great expense; and here, almost the whole mill and machinery must be improvised of wood, on the spot. Timber, water power, and space being of little value, while *iron* is a precious metal, and he must exert his ingenuity to construct a mill that will do the required work with the greatest possible economy of iron work. Mills are sometimes thus built almost entirely of wood, without bands or bolts, or even nails, with scarcely anything metallic except the saw.

These supposed saw-mills occupy the two extreme situations, but every location upon which a mill is to be built involves a modification of material and construction continually varying according to circumstances, requiring the millwright to determine and select, with a due regard to contingencies, the kind of wheels, and machinery, and the materials most suitable to the site. How, and of what material to construct the dam? is a part of the problem continually changing with the situation.

These facts tend to show that a "boss" millwright has need of an extra share of tact and ingenuity, which will be obvious when we consider that a like diversity of conditions occurs in constructing other mills and machinery, as well as saw-mills. The following considerations will further illustrate the head millwright's responsibilities: One employer has a small stream of water with a high head; another has a large stream, with little or perhaps no chance of raising any head above the natural current; another has a good water power, and well situated, but very little means to lay out upon it. The next has plenty of money, but a water power and situation almost impracticable; each wants some kind of mill built, and the millwright must plan and construct a suitable mill for any one, or all of these.

Any man who would make a good ordinary mechanic may learn the use of tools, and make a good journeyman millwright, but a thoroughly competent master millwright must, like a poet, be born such, and cannot be made. Hence it is that we find so many celebrated millwrights who never served an apprenticeship to that trade, but were originally perhaps a miller or sawyer, a carpenter and joiner, an engineer or machinist, until some accidental circumstance occurred to show that they were millwrights.

Such men come into the trade naturally by intuition, but even they would acquire the requisite knowledge much sooner, easier, and better, by serving a term of apprenticeship under an old experienced master, doing an extensive business, than having to study and work out every problem for themselves. Observation shows that the more information and experience a millwright obtains, the more apt and ready he becomes to learn; and he can scarcely examine any old mill, or converse with any old millwright or miller without obtaining some new idea, or perfecting some old one. A full and free interchange of ideas and experience among leading millwrights would be of great advantage to the faternity, but of this they are naturally chary, for the reason that each one has had to work out so many important problems for himself, that he looks upon these discoveries as his own exclusive property, and will leave others to work them out for themselves as he had to do, or remain in ignorance.

This reticence confines the aspiring learner mostly to his own resources, and the help obtained from books; and experience has taught us how meagre and unreliable most of the works on the subject are; what a small amount of real practical information they contain, tedious to pick out, and unsatisfactory when obtained.

Some scientific books contain theories and formulæ by which the velocity and power of water under different circumstances may be computed; but these, like the philosophical treatises on machinery, are generally written in a style that is not very intelligible to the ordinary learner, who, after he has unravelled all the mysterious scientific terms and algebraic formulæ in which they are enveloped, often finds that they are inapplicable to his case, or unreliable, and he finds it easier and safer to gather information by studying working machinery, conversing with individuals who understand these subjects, or by experimenting with water and other things himself, and working out problems with the square and compasses on a planed board. The work published by Oliver Evans was an excellent practical work in its day, but mills have so changed within the century that it is too far behind the times to be of much use at the present day.

We have frequently been requested to write a book, embodying our own long and rather extensive experience in building and running different kinds of mills in many sections of this country, and what knowledge we have managed to pick up during that career, for the benefit of others similarly employed or interested; these requests being backed by our own knowledge of the want of such a work, from the many fruitless efforts to procure one, have induced us to make the attempt. How far we have succeeded in supplying the want, the reader will judge. The book makes no claim to literary merit further than to convey the knowledge and experience gathered in the career of a hard-working practical millwright and miller, in plain concise terms, without any algebraic or scientific mystery, and is wholly original. Some of the chapters, as those on the manufacture of oatmeal, pot and pearl barley, and split peas, that on windmills, and some others, are subjects of which comparatively little is known in the United States, and our knowledge of these subjects was gathered during an extensive experience with these mills in the adjoining dominion of Canada and in northern New York where such machinery is more common.

We have been assisted in the arrangement of the present work by Mr. Robert Middlemiss, of Rockburn Prov. of Quebec, Canada, who has also assisted in preparing the drawings. The general scope of the work is to give practical and reliable information to millwrights and millers generally, but more especially to that large class who pursue their vocation in localities distant from large machine-shops; also to persons contemplating erecting mills and machinery in such situations; and the author trusts that such will find much information in it which they daily require, collected and compressed within a smaller compass than in any work of a similar nature. The tables will be found reliable and so simplified as to be easily understood by all. To the young aspiring millwright and miller especially, we would say, earnestly endeavor to improve your mind, and enliven your spare hours by the cultivation of practical science; and should this volume facilitate your progress, the desire of the author will be realized.

DAVID CRAIK.

CHURCH MILLS, CHATEAUGAY, FRANKLIN CO., N. Y.,
June 20, 1870.

THE PRACTICAL

AMERICAN MILLWRIGHT AND MILLER.

CHAPTER I.

MECHANICAL POWERS.

Elementary works on natural philosophy, treating upon the subject under the head of mechanics, mechanical powers, or elements of machinery, are so numerous, and so widely distributed, that almost every person who may be expected to take any interest in reading this work, must have seen and read more or less of these, and we will, therefore, say but little on this subject.

One reason for this reticence, we will confess, is a sort of repugnance, which we, in common with other millwrights of our acquaintance, feel in wading through the hackneyed and bewildering technicalities which have been introduced into it. This is the result of the many unnecessary divisions and subdivisions which the various authors have seen fit to make out of the lever and the inclined plane. One author copying from another, and intent on showing his greater erudition by his nicer discrimination (while very few of them were practically familiar with the instruments they were explaining), and each new writer felt bound to add something new to the elucidations of his predecessors. We have a new work before us now, which, from the position the author assumes, ought to be reliable, as he writes, like us, expressly for the information of millwrights; yet his explanation of the principle and power of the lever seems to us more glaringly absurd than any we have seen by a mere theorist. For these reasons, millwrights, in general, look upon all disquisitions on these subjects as interesting only to amateurs, and not available as a basis for practical calculations.

We once undertook, in a bantering way, to raise a stick of timber to its place by merely walking upon it; all the help we asked being a boy to block it up when we raised it. The stick was forty-five feet long, and had been run into the basement of a mill, and lay angling from one corner to the opposite one, upon two blocks, three or four feet apart, near the centre. It lay two or three feet from the floor, and had to be raised four or five feet higher, and then swung parallel with the building, when the two ends would reach and rest upon the end walls. A pile of stuff cut for buckets lay convenient, and a pile of 3 by 4 inch braces. We placed a bucket piece on each side of the timber, these reached from one to the other of the timbers supporting the stick to be raised, and made the commencement of the cob-house or crib work upon which to raise and support it. We now stepped on the stick and walked up to the most elevated end; this brought that end down, throwing the weight all upon the bearing nearest to it, at the same time elevating the other end, and making room for the boy to place a brace across the bearing next to that end, which kept it up. By walking up along the stick to the now elevated end, that end was depressed, relieving the other bearing, and making room for a brace to be placed across upon that. By placing two more bucket boards across these braces, and continuing the process, the stick was raised to the height required, in less time than we can write the details of the process. The last bearing was placed across the middle of the cob-house at the balance of the stick, and upon this it was easily swung round to its position on the walls.

As we are writing for the benefit of millwrights, this description, without any accompanying draft, is all that is necessary, as they will comprehend the principle involved at once, and be satisfied. But if we were writing for the small fry philosophers alluded to, we would have to accompany that description with a plan, and a tedious and intricate formula, to demonstrate the principle of the *lever* upon which our weight acted; another to demonstrate the principle of the inclined plane, up which we travelled, carrying the stick (weight) up behind us; and yet another to show that the principle of the balance was as much involved as either of the others; and the most inveterate sticklers for points and principles would have it a tread power, or expect us to show that each piece of scantling (brace) used, represented a wedge, and that each lift was accomplished by the intervention of a wedge, with the head and point reversed each time alternately.

After making this long preface and these ill-natured remarks, it might not be expected that we should tell how many kinds of levers there are, nor point out their distinct peculiarities, but as the principle of the lever in some shape or form appears in almost every machine, however simple or complex, it is important that its various combinations and relative powers should be well understood. We will, therefore, try to point out some of its most obvious applications, and thus illustrate its effect in increasing or diminishing power, or of increasing or diminishing velocity.

The Lever.

The handle of a common pump is a familiar example of a lever. The weight to be lifted is the pump-rod and valve, and the column of water. The fulcrum or point of support is the pin through the handle, and the power is the hand applied at the long end to work it. If the pump-rod is attached at one foot from the fulcrum, and the hand applied at four feet from that point; then every pound pressure by the hand will raise four pounds on the pump-rod, but the hand will have to move four feet down and up to move the pump-rod one foot. Fig. 1.

Fig. 1.

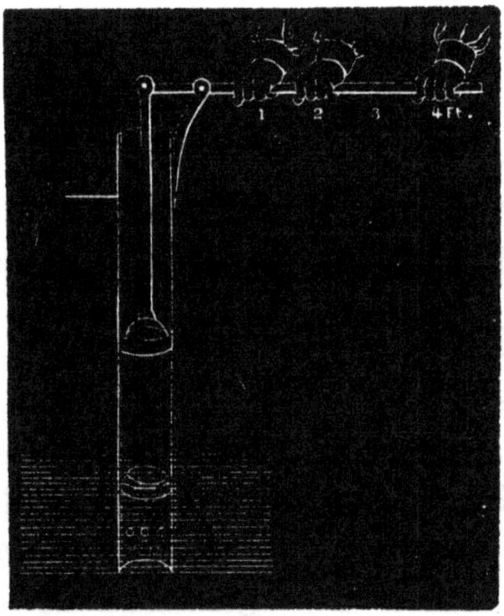

If the hand be applied at two feet from the fulcrum, then every pound pressed by it will balance two pounds at the pump-rod, and the hand will have to travel two feet to move the pump-rod one. If the hand is applied at one foot from the fulcrum, the same as the pump-rod, then a pound upon one will just balance a pound on the other, and to move the hand one foot, or an inch, will move the rod exactly a foot or an inch. Both arms being equal in length,—power and velocity will both be transmitted without either loss or gain from either end of the lever.

If the hand is applied at only half a foot from the fulcrum, it loses the advantage in power, but gains in that of speed, the relative proportions in the length of the lever being reversed; and it will now take two pounds pressure by the hand to balance one pound at the pump-rod,—but a motion up and down by the hand of half a foot, will move the pump-rod up and down a whole foot.

Here it must be particularly noted, that the tables are now turned in favor of speed, and against power, the pump-rod being on the long end of the lever, and the hand on the short end, and that every increase in the difference of length in this direction, gives the pump-rod an increased length of stroke, with a corresponding diminution of force, while it requires an increased force to be communicated by the hand, but with a corresponding diminution of the distance through which it moves, and that this arrangement gains speed by wasting power, just as the first arrangement gains power by wasting speed. The point at which neither speed nor power is wasted or gained being where both ends of the lever are of equal length; and here the power has the same effect as if applied by the hand taking direct hold of the pump-rod, and all the advantage derived from a lever thus situated is, that it transposes the downward pressure of the hand into an upward lift on the pump-rod, thus enabling the hand to work in unison with gravity, instead of opposed to it.

It is the benefit derived from working in the direction

of gravitation that has established the use of this kind of lever for lift-pumps, blacksmith's bellows, and other machines requiring an upward lift, rather than the second kind of lever, which is used for force-pumps and other machines which admit of a downward pressure.

That this second kind of lever is more powerful than the first, may be seen by examining the action of the same pump lever that we have been trying to illustrate. Supposing the column of water to be raised, with the pump-rod and valve, to weigh 40 lbs., the lever, as before, being one foot from the fulcrum to the pump-rod, then it would require a pressure of 10 lbs., by the hand, applied on the long end of the lever (handle), at four feet from the fulcrum, to balance the 40 lbs. The 10 lbs. and the 40 lbs. being both supported by the fulcrum, make 50 lbs. pressure upon it. If the pressure be applied on the lever at one foot from the fulcrum, then both ends being equal, it will require 40 lbs. pressure by the hand to balance the 40 lbs. on the pump-rod and water, thus making 80 lbs. pressure on the fulcrum, or double the amount transmitted through the lever to the other end.

Now, by moving the fulcrum bearing to a foot behind the pump-rod, and attaching the rod to the lever at the former fulcrum point, as in Fig. 2, the lever is transformed to one of the second kind, the fulcrum being at the end, and the rod or weight attached a foot from that point, a pressure up or down upon the long end will now produce the same increased force upon the pump-rod, that it exerted before upon the fulcrum, *i. e.*, the sum of both weight and pressure, instead of the equivalent of one. This apparent gain of power by the lever has been the cause of many extravagant and expensive errors. Some millwrights increase the number of intermediate parts in a machine, expressly to gain this imaginary advantage.

Other mechanics and geniuses will complicate and confuse the simple appliances required to drive a churn or washing

Fig. 2.

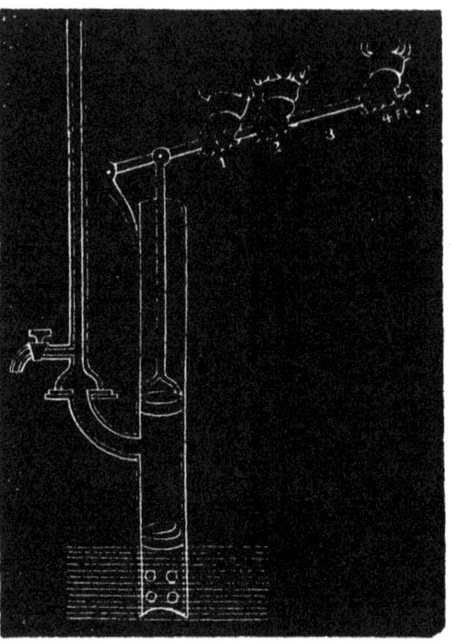

machine for the same reason. The opinion is quite prevalent, that when a person holding the suspending hook of a steelyard, Fig. 3, lifts a weight, say of a 100 lbs., on

Fig. 3.

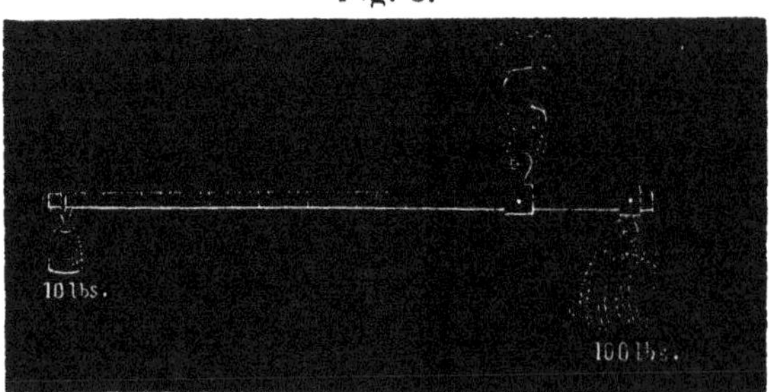

the hook at the short end of the lever, he must sustain a weight of 200 lbs., because, they say, the long lever power on which the steelyard ball acts, enables it to pry downward on the fulcrum support with a force of 100 lbs., being equal to, and balancing that on the short end. The most extensive application of this fallacy that we have ever seen was among the farmers in some of the Middle States, particularly in the southwest portions of New York. The ploughing is often done when the land is dry, and too hard for one pair of horses, and the custom is to attach three horses to the plough, all abreast. The two strongest are hitched to a common evener or double-tree in the ordinary way; the centre of this evener is attached to the short end of a long lever or triple-tree,—this is fastened to the plough clevis at one third of its length from the double-tree end, leaving two-thirds of its length for the long end, to which the single horse is attached. The arguments by which they support this arrangement are, that by taking advantage of the gain in power, by the extra purchase of lever on which the single horse acts, its power balances, and is equal to the other two; and thus, they say, they have the power of four horses on the plough, with only the expense and trouble of managing three. The principle involved is exactly the same as that in the steelyard referred to; and the fictitious gain vanishes when we consider that there are only the 100 lbs. of weight on the short end, and the ball on the long end, with the weight of the intervening bar and its fixings, say 10 lbs. more, added to the weight, making altogether 110 lbs. balanced on the fulcrum, and held up by the person holding the hook. If this is not sufficiently plain, move the fulcrum and hook to the middle of the bar, as in Fig. 4, and divide the weight into two, of 55 lbs. each, and hang one on each end of the bar, or divide the respective lengths of the lever in any other proportion, apportioning the weight in the same ratio, and the weight upon the suspending hook will always be the same, at whatever point of the bar it may be balanced. The same remarks ap-

Fig. 4.

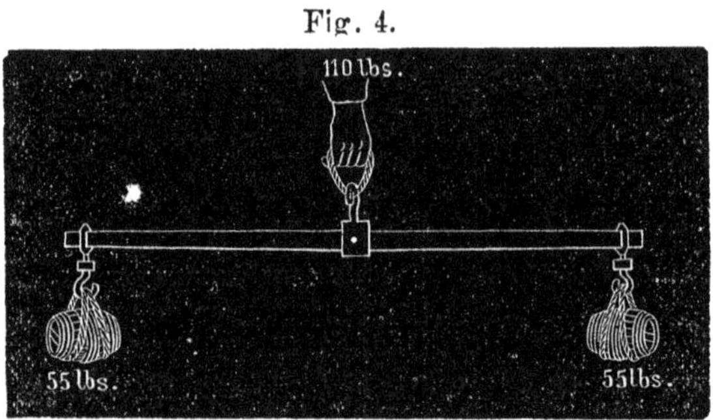

ply to the other problem of attaching three horses to a plough, as there is no principle in the lever or any other power in mechanics, by which their actual power and speed can be augmented.

We have already seen how a *simple* lever may be used to increase power at the expense of speed, or to increase speed at the expense of power, but by a *combination* of levers more or less complex these transpositions may be made in endless variety, and to an almost unlimited extent. The principles involved in these combinations are so nearly alike, and at the same time so obvious, that they reveal themselves in the construction or planning, or even in the examination or contemplation of machinery, much better then we could by writing of them. Therefore it will only be necessary to notice a few of the most important points to be kept in view by the millwright, and to illustrate these we will go back again to the pump. But this time, instead of locating the fulcrum either before or behind the pump-rod, we will fasten it exactly over the centre, as shown in Fig. 5. In the last experiment detailed the power was applied on the lever only half a foot from the fulcrum. We will now insert a handle or pin at this point and attach the pump-rod to the same pin; now cut off the lever outside of the pin, and, by taking hold of the pin, revolve it round and round like a crank. This will carry the end of the pump-rod round a circle one foot in diameter, and lift and depress the valve-bucket one foot each revolution, the same as in the other examples with the lever; but in this case, without either loss or gain of power or speed by the lever, its only effect being to steady the motion of the hand and determine the length of stroke. But if we separate the power from the end of the pump-rod and attach each by a crank of its own to the opposite ends of a short shaft, then we can lengthen the crank end to which the hand is applied and gain power on the pump, or shorten the same crank and gain speed on the pump, its length of crank and stroke remaining the same; or, we can shorten the crank and stroke of the pump-rod and gain power, or lengthen these and gain speed in it—the length of the hand crank remaining the same, as in Figs. 6 and 7. Then we can combine these alterations in both directions and thus double the loss or gain made by either. So far these variations are all made by the use of a single lever by altering the respective lengths of the two ends. If we wish to gain more power or speed than this admits of then we must have recourse to a combi-

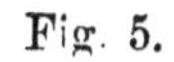

Fig. 5.

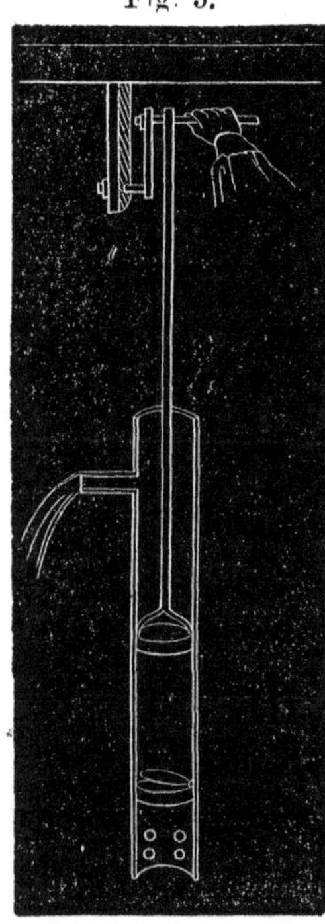

nation of two or more levers. To apply these, the crank to which the pump-rod is attached must be changed into

Fig. 6. Fig. 7.

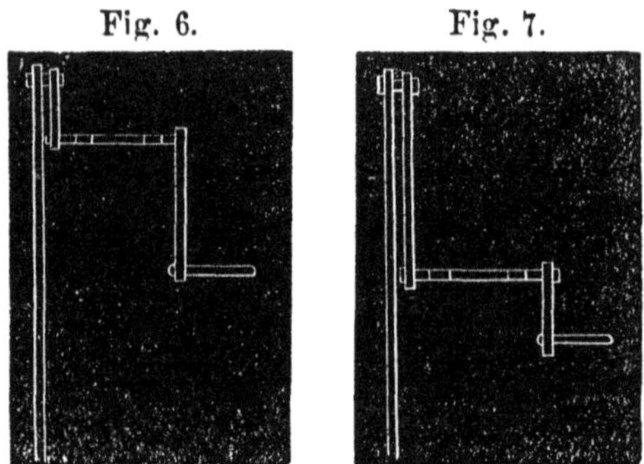

or furnished with a wheel (a continuous revolving lever). The crank to which the hand or power is applied must also be transformed into a wheel and hung upon a separate shaft. The circumference of these wheels must be connected by cogs, or a band, so that the motion and power of one can be communicated to the other, as in Figs. 8 and 9.

To gain *power* by these, the wheel to which the pump is attached must be enlarged, *i. e.*, its lever power must be lengthened, and the wheel to which the driving-crank or power is applied must be diminished, thus shortening its leverage and giving it the advantage. To gain *speed* and lose power it is obvious that the enlarged and diminished wheels must change places. When it is necessary to gain more power or speed than two wheels will admit of, then one or more wheels, according to circumstances, must be added—thus forming a train of wheels in which each pair gains a proportion until the required result is attained.

The movements in a clock furnish an example of a train of wheels gaining speed from a slow motive power. A better example for a millwright will be a large overshot

Fig. 9.

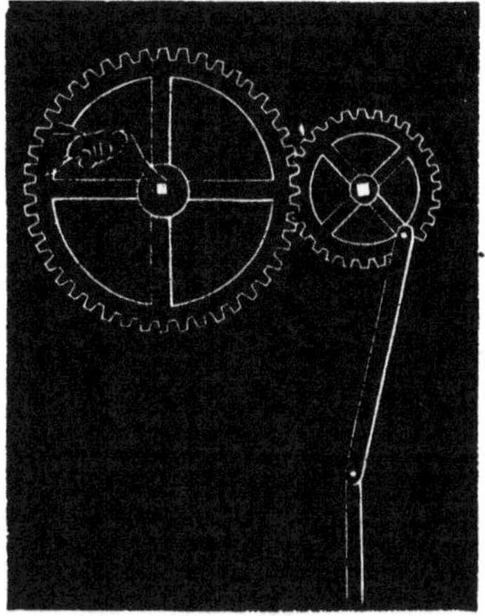

wheel making only three or four revolutions in a minute geared on to a circular saw making fifteen hundred re-

Fig. 8.

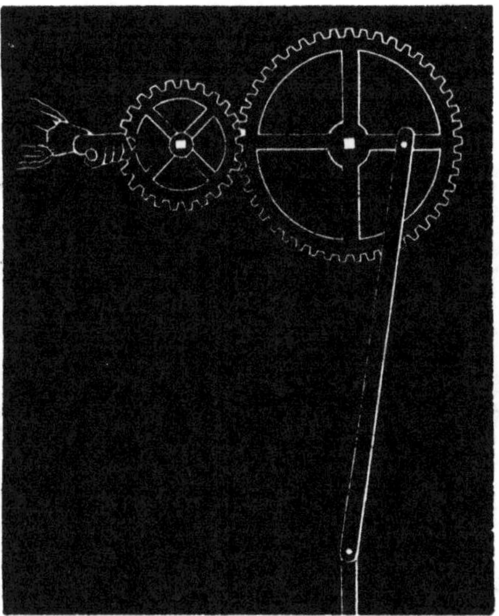

volutions in the same time, and one of Leffel's six or eight inch turbines, making fifteen hundred revolutions per minute, geared to some slow moving heavy machine.

A millwright should avoid, as much as practicable, such extremes, because the less machinery that is interposed between the power and the work the better the result will be. It is cheaper in the first place, easier worked, and less liable to get out of repair.

The Inclined Plane.

The inclined plane is a mechanical power so commonly used, so simple in principle and varied in its applications, that people often make use of it in their daily avocations without knowing or suspecting that they are dependent upon any of the (so-called) mechanical powers for assistance. The teamster transports his load from the valley to the summit level, or up the mountain, by a natural inclined plane improved into a road. The backwoodsman rolls the logs on to his sled, a log heap, or a log house, by means of skids, another inclined plane. The merchant loads or unloads

Fig. 10.

his barrels and hogsheads, or transfers these into or out of his cellar by a plank or two timbers placed parallel, which is also an inclined plane, Fig. 10. In digging a canal or raising an embankment the earth is wheeled up an inclined plane on the same principle.

In all these and similar operations the weight or load is raised to the required height by the continued application of a force, the power of which is somewhere between that required to move the load along a horizontal plane and that which would lift the whole load bodily and perpendicular. For this reason the method of expressing the angle of inclination by the perpendicular rise in a certain distance, as one foot in five or one foot in ten, is preferable to that of expressing it by the degrees in a quadrant. This will be evident when we consider that the power required to move the load up a plane is always in the ratio of the perpendicular rise to the horizontal length of the plane. That is, a rise of one foot in four will require a force equal to one-

Fig. 11.

fourth of the weight of the load; one foot rise in six or ten will take one-sixth or one-tenth of the weight of the load to move it up the plane. This, like all other theoretical estimates, ignores the resistance of friction. A knowledge of this law of inclined planes is necessary to the mechanic, as he has often to adapt machines to work on different inclines, such as tramways connected with mines or public works, where a carriage descending one plane is made to draw up another carriage on another plane, as in Fig. 11.

In the instances referred to the load is carried up the plane by the continued application of a *part* of the force required to lift it. But there is another application of the inclined plane up which the load is carried by steps or instalments, the *whole* weight being raised by a succession of short lifts from one resting point to another. Stairs leading up into the higher stories of buildings, or down into cellars, are familiar examples; or the inclined side of a pyramid up which the materials for its construction are conveyed by a succession of short lifts from one course of masonry to the next one. Various mechanical contrivances, more or less simple, are employed in connection with the inclined plane for raising or lowering heavy materials (it is equally applicable to both). Such are handspikes or ropes and pulleys used in rolling up or letting down heavy logs and barrels, also the jackscrews and powerful windlasses employed in launching vessels or hauling them up into docks. It may be remarked that the gain of power by the inclined plane, like that gained by the lever, is always at the expense of time or speed. Suppose a loaded carriage has to be raised on to a floor one foot in perpendicular height above the level road, we may fix one incline to extend back ten feet from the floor, another to extend twenty feet, and another forty. It will require the *same power* to draw the carriage over the forty feet, with one foot rise, as to draw it over the twenty or ten feet, with the same one foot rise, and that will be exactly the power that would raise the carriage one foot perpendicular.

This shows that there is no power really gained by the inclined plane; it is only a medium of accommodation by which a fortieth part of the power required to lift the carriage one foot high, by being exerted over forty times the space and time, is enabled to lift the same weight to the same height. It may be remarked, that the laws governing the ratio of power and effect are not arbitrary, as the last statement might imply, but the power is increased as the horizontal length of the plane is increased, and diminished as the perpendicular height of the plane is increased.

The Wedge.

The wedge is another modification of the principle of the inclined plane. In those already described the weight or load is raised by being forced up a stationary incline from the base or point to the head. With the wedge this order is reversed, the base of the plane or point of the wedge being inserted under the load to be raised or between two objects to be forced asunder, it is then forcibly driven to the head. By this it is to be seen that the distance to which anything can be moved or raised by the wedge is very limited. But this peculiarity, which limits its effective range to so narrow a space, gives it at the same time some valuable advantages, enabling it to be used in cramped situations and narrow spaces which would not admit of the lever or screw, or any other mechanical applications. Under such circumstances it is not only the most convenient mechanical contrivance that can be used, but also the most powerful. Suitable wedges made of wood or iron, and driven between substantial bearings by a heavy maul or a battering-ram, exert a force sufficient to raise almost any amount of weight, or burst asunder almost any material. We have seen a crown-wheel or pinion weighing half a ton bursted open by wedging it with wooden wedges upon a wooden shaft. Similar wedges of iron or steel driven into a fissure, or into holes drilled for their admission in a rock, will burst the rock asunder like the pinion.

The following method of dividing large masses of stone for making mill-stones was formerly practised in Scotland and Germany: the block, after being broken from the rock, was dressed off round in the form of a pillar, the thickness of which was equal to the intended diameter of the mill-stones; the pillar was then divided and marked into sections, each being the intended thickness of the stone. Holes were now drilled in these marks all the way round the pillar, and *dry* wooden wedges were driven tight into these holes; the wedges were then *wet*, and the gradual and equal expansion of the wood, by the absorption of the water, rent the sections asunder. This might, of course, be accomplished by placing two half-round iron feathers, with the thick end down, in each hole, and then driving steel plug wedges equally all around between these feathers; this is the process by which building stones are wedged apart, but the effect is not so equal, and is liable to injure the texture of the stone opposite the wedges too much for mill-stones.

Grindstones are sometimes made in this way, but only in such quarries as do not lie in seams or layers. When found in this condition thin steel wedges are driven in the seams or fissures dividing the strata, which are thus raised from the ledge in large flags; these are then marked for stones of various diameters to avoid waste of materials, broken apart by plugs and feathers or otherwise, and then dressed off to the circle, the depth of the strata or thickness of the flags, making the thickness of the grindstones.

Another example of the convenient application of the wedge is the use made of it in fastening the handles into hammers, axes, and all such tools, and also in fastening the gudgeons, cranks, or other such irons into wooden shafts. In this connection we might likewise mention its application in tightening and fastening bands, pinions, wheels, couplings, and all such attachments to shafting, whether by ordinary wedges or square and tapering keys. For these and all other similar purposes there is no substitute for the wedge known that would be equally applicable and effective.

This principle of the wedge renders very effective aid in using the axe, the chisel, the drawshave, and many other tools, by forcing the material apart and loosening the texture in advance of the cutting edge, thus assisting its entrance. It may also be observed that the wedge is frequently assisted in its entrance in a similar manner by the principle of the *lever*, thus the seam of the rock is pried open in advance of the wedge by a leverage equal to the distance of the fissure and the strength of strata, and the timber is forced apart in advance of the wedge in the same way, in the process of splitting a log.

The Screw.

The screw is another modification of the principle of the inclined plane, but its construction and application are quite different from that of the wedge, which, although used in so many different forms and situations in the construction and arrangement of machinery, furnishes no medium of transmitting or communicating continuous motion. The screw can be modified to produce a dead lift, like the wedge, or to communicate almost every variety of motion: rectilinear, reciprocating, or rotary.

The most common and obvious application of the principle of the screw is, perhaps, the screw-press, Fig. 12, or the ordinary jack-screw for raising buildings, Fig. 13.

Fig. 12.

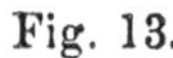

Fig. 13.

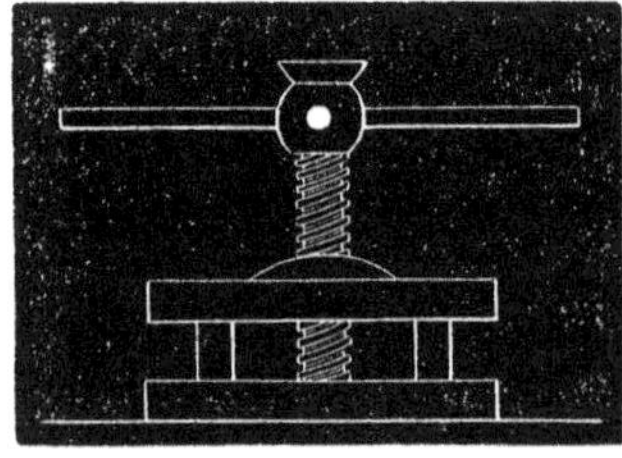

Other and still more common applications are the screw-bolts and nuts used for fastening almost every description of structure together, and screw-nails. The screw, in all its various modifications, is a combination of the two primitive mechanical powers, the inclined plane and the lever. The *thread* of the screw is a continuous inclined plane wound spirally round a central shaft or axis. The wrench or bar by which it is worked is a lever, the fulcrum or axis of which is the central shaft of the screw; the power of this lever is found by dividing its length from the centre of the screw to the point where the power is applied by one-half of the diameter of the screw. For an example, suppose a screw be four inches in diameter and the lever three feet long, then divide the length of lever, 36 inches by 2, the half diameter of the screw would give 18, the number of times the power applied is increased. The power of the screw is found by dividing the length of one coil of the screw by the distance from the top of one thread to the top of the next thread above it. Here we will call the length of one revolution of the coil 12 inches, and the distance of pitch half an inch, then we have $12 \div \frac{1}{2} = 24$, being the number of times the power applied by the wrench (lever) is multiplied by the screw (inclined plane). Then $18 \times 24 = 432$, the number of times that the power applied to the wrench is increased by the combined action of the machine.

It may be necessary to add that the naked screw has no power of itself; it can only operate by pressing against the thread of another screw which overlaps it, and is called a box or nut; it consists of a block with a central tube cut out in spiral grooves so as to fit with the screw which has to work in it. The lever is applied either upon this nut or through an enlarged head upon the screw itself.

The screw, like all other mechanical powers, loses in speed or distance all that it gains in power. The distance between the threads being diminished increases the power as the incline is diminished; the distance of pitch being increased, increases the speed as the incline is increased. To make this plainer, suppose two screws, otherwise alike, but one an inch pitch of thread, the other half an inch pitch. Two men would raise a ton weight an inch high by one revolution of the lever of the first, while by the other screw one man would raise the same weight to the same height by two revolutions of the lever.

Practice and observation have established approximate rules for graduating the pitch of thread, thickness of nut, &c., to the diameter of the screw and purpose to which it is to be applied, but these would be out of place in a work like this.

The screw used for feeding the logs through a slabbing gang of saws is an example of a screw producing a continued forward motion. Those used for moving the head-blocks and dogs of circular and other saw-mills give a motion both forward and backward. The endless or everlasting screw communicates a continuous revolving motion to machinery. This is frequently used in mills to obtain a slow motion from a swift revolving shaft or spindle. The most familiar example of this device is the ordinary clock reel used for winding yarn into skeins, and counting the requisite number of forty threads. This is done by cutting a single thread screw around the axle of the reel and fitting the screw to work in the edge of a small wheel, the circumference of which is equally divided and cut out into forty notches or cogs; these cogs, as also the thread of the screw, are cut conical or V-shaped to an edge, and thus match into each other. By turning the reel, each turn of the thread upon its axle receives a tooth of the wheel and brings it forward, and as soon as one turn of the thread is disengaged another comes into operation, so as to produce a perpetual revolution of the wheel; a short pin is fastened into one side of the wheel near the edge, and catches a spring which it bends until the end slips off the pin and the spring reacts back with a snap, which announces that the wheel has made its full revolution, and, consequently, the reel its forty revolutions, and therefore contains the requisite forty threads, which are now tied and the operation repeated.

By combining the screw with the *wheel* and *axle* in this manner a slow motion is obtained, which is very strong and uniform, and is generated by a very slight power, but this, of course, must have a corresponding swiftness of motion to compensate.

The wheel and axle here mentioned is shown in Fig. 14; it is identical with the wheel and pinion, the wind-

Fig. 14.

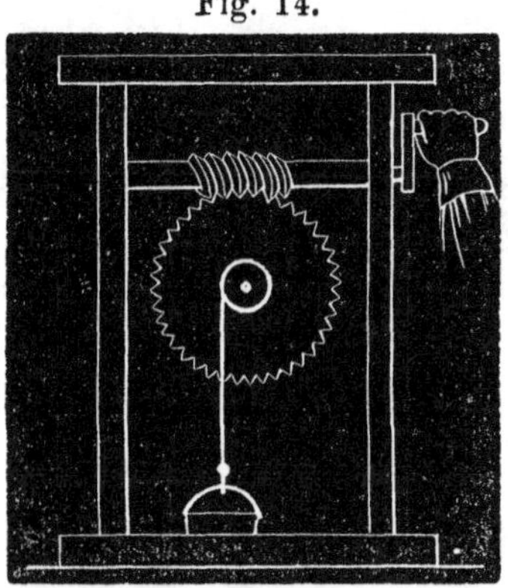

lass, and similar contrivances, all of which are revolving or perpetual levers; the wheel or crank being the long end, and the pinion or the shaft upon which the rope or chain winds, as the case may be, is the short end of the same lever. The gain of power or of speed by any or all of these is always in the proportion of the length of these ends; *i.e.*, the relative diameter of the one to which the power is applied, and that upon which it produces the effect.

Examples of the application of the principles combined in these machines are numerous and common. The wheel and axle is used to hoist goods up to the higher stories of store-houses, or to raise coals and other minerals from mines. It is also combined with other wheels and pinions to increase the power of men applied to cranks, as seen in the crane, Fig. 15.

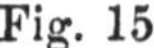

Fig. 15

The *bull*-wheel or *pull*-wheel used for drawing the logs up into saw-mills is an application of this principle with which millwrights are familiar.

These are all examples of gaining power at the expense of speed, but the same principle is applied with equal facility in the opposite direction as already intimated; that is, to gain speed by the expense of power. Take any of the machines alluded to, as the crane for an example. When the weight has been raised to the full height, if the men should let the winch or crank go, it would turn back in the opposite direction with great and increasing velocity until the weight had reached the point from which it was raised, thus giving back the same number of revolutions and the same amount of power which it took to raise it. This is the order of gearing employed for lathes, circular saws, &c., when driven from a slow and strong first mover.

There are many other mechanical devices in use; those mentioned are either for gaining power or speed, but there are others employed to change one kind of motion into another, as cranks, eccentrics, and cam-wheels, for changing rotary into reciprocating, or *vice versâ*, also universal or toggle joints, bevel gearing, &c., for changing the direction of revolving motions.

The Pulley and Cord.

The pulley and cord is another mechanical contrivance. The fixed pulley, Figs. 16 and 17, gives no increase of power, its only advantages are that it diminishes friction and gives convenience in pulling. The movable pulley doubles the power, as may be seen by

Fig. 16. Fig. 17. Fig. 18. Fig. 19.

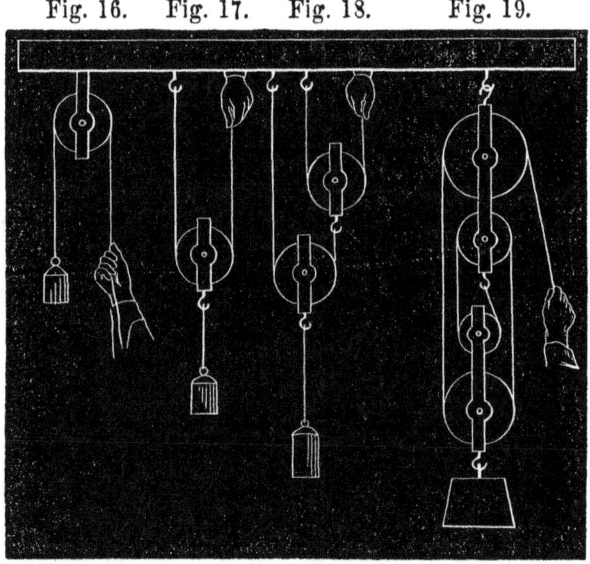

Fig. 18; one end of the cord being fastened, the power applied upon the other end exerts a double force upon the pulley at the middle by losing one-half of the speed.

The blocks and tackle used on shipboard and in raising buildings, also for hoisting heavy materials, are systems of fast and movable pulleys; the fast ones are contained in the upper block, and the movable ones in the block attached to the weight to be raised. The concentration of power by these combinations is as the number of pulleys in each, and is determined by the number of movable pulleys, each one of which doubles the power, therefore multiply the number of these by two and that product by the power applied: this last product is the weight the tackle will raise, or the full measure of its accumulated force.

It will be seen by this that the addition of one movable pulley to any system of pulleys doubles the effective power, but at the same time it also doubles the time required to raise the weight, as every additional fold of the rope makes so much more rope to draw out, Fig. 19. By fixing a windlass or wheel and axle to wind up the rope of a block and tackle, the power of the lever is combined with that of the block and tackle, and there is scarcely any limit to the power which may be exerted by such a combination, except the strength of the materials used.

The pulley upon which the cord acts in these systems is merely a continuous lever with equal arms, and giving no mechanical advantage. It is the equal distribution of the weight among all the folds or strands of rope which divides and suspends it upon all the intervening centres of motion that gives this advantage. Levers with arms of *unequal* lengths are frequently used, sometimes for gaining power, in other circumstances for gaining in length or distance of range, and also for changing the direction of motion, and similarly worked by a cord. Instances of this kind of cord and lever occur in mills where they are employed to regulate the feed, throw tighteners off or on to belts, or to manipulate the traps and gates for distributing grain from elevators, &c. Such arrangements are called, by learned authors, animal levers.

The *power* or *velocity* which these either gain or lose is estimated by the respective lengths of the two ends, measured from the fulcrum or centre of motion, like all the other levers heretofore mentioned and referred to; but when continuous revolving levers or wheels are used the gain or loss is much easier computed by measuring their relative circumferences. If cog-wheels, count the number of cogs in each and divide the larger by the smaller; if belt-wheels or pulleys be used, then measure round their circumferences with a tape-line or other string, taking the girth in inches, and divide the larger by the smaller, the same as with cogs.

The Crank.

Cranks, Fig. 20, are another kind of revolving levers, whether they are used to transform a reciprocating

Fig. 20.

motion into a rotary one, as in the steam engine, or to transpose a rotary into a reciprocating, as in the common saw-mill.

This lever differs from all the others we have been considering in having the peculiarity of gradually varying its length of fulcrum from the whole length of its radius down to nothing, and back again to full length twice during each revolution, and it maintains this uniform gradation of length in both the situations referred to: that is, whether the reciprocating motion be applied to revolve the crank, or the crank be employed to produce the reciprocating motion. An exception to this may be thought to be found in the application of the hand, or other revolving motor that accompanies the wrist-pin of the crank in its circuit; but such is not the case, because when thus situated it ceases to be a crank, and is only an ordinary revolving lever—the wrist-pin being only the medium of connection, like any other coupling.

There is nothing that we can think of in the whole range of mechanics that appears so simple, and yet with regard to which so much misconception exists, as the crank. And this is the result of the tendency which exists in the ordinary mind to see and appreciate the advantage of power gained by lengthening a lever, and ignore the loss in time and space which that gain involves. To dispel this illusion, and banish the prejudice against the crank, it is only necessary to consider that the steam or other power applied in a right line upon a crank must follow, and be exhausted, just twice as fast at the middle of the stroke, where the crank acts as a lever of its own length, as at half way to the end, where it acts as a lever of only half that length. So that the crank, like every other mechanical device, maintains always the same ratio of power and effect with time and space.

Another delusion is very common with regard to the crank, which is, that by making the iron arm of the crank considerably longer than the intended length of stroke, and then crook it up in the form of a half circle or the letter S, a great advantage in power is gained. We remember hearing this point argued, when very young, in connection with the cranks used for turning the hand-mills for grinding coffee and spices. But in the year 1836, while employed with the engineers in laying out the works for the renewal of Fort Mifflin on the Delaware River, a very amusing controversy occurred on this subject. The *moats* or canals around the fort had to be pumped dry, and the screw-pumps used for that purpose were worked by six or eight men each. A bevel wheel on an inclined pump, geared into another on a horizontal shaft, fixed across a supporting frame: this shaft had a crank upon each end, each crank being worked by three or four men. The cranks were of wrought iron, and made to get the full benefit of this *advantageous* leverage; whether from the simple conviction of the maker or to humor the popular prejudice is not material—but the cranks were continually breaking and causing interruptions; and as there was no heavier iron on the works, nor nearer than Philadelphia, the question arose, whether the arm of the crank might not be shortened one-half by making it straight from the shaft to the handle. It was admitted that if this were done the same iron would be strong enough to hold the men, as only four could get hold of the handle at the same time; but some argued that the handles would have to be lengthened to allow more men to get hold before they could work the pump, and then the crank would break as before; and these could not be convinced until they saw the straight crank applied, and the pump worked by the same hands and with the same ease as before.

Some will think such ideas too whimsical and frivolous to mention here, but they are often troublesome and mischievous as well as whimsical; and we have been so often annoyed with the same kind of philosophy since, that we think the mention here made may induce some individuals to review their "theory of the bent lever," and thus be the means of getting this particular kink out of their imaginations.

The bent cranks used for tread powers, such as those driving turning-lathes, grindstones, or the old-fashioned spinning-wheel by the foot, are instances of the prevalence of this chimera of lengthening the lever, and at the same time maintaining a short range of stroke.

The eccentric, so common in steam-engines and other machines, is a modification of the crank, and its principles and mode of action are identical with those of the crank. It is used where the situation or size of shaft or some other peculiarity does not admit of an ordinary crank. The various descriptions of cams are also modifications of the eccentric. The crank and eccentric have but one back and forward motion for each revolution of the shaft to which they belong, while by cams two or more may be made, as in driving hide or cloth fulling-mills, trip-hammers, and also many other reciprocating movements in carding-machines, &c.

CHAPTER II.

FLY AND BALANCE-WHEELS.

THE fly-wheel and the balance-wheel, and their different uses and applications, the millwright should also be familiar with. The fly is used to equalize velocity, the balance to equalize weight.

When a crank is used as the medium of transmission between a rotary and reciprocating motion, in either direction, a fly-wheel is generally necessary to equalize the momentum among the various situations of the crank during its revolution. The size and weight of the fly must, of course, be determined by the weight, and more particularly by the nature of the machinery. In many instances this contains a sufficiency of regulating influences within itself, as the locomotive engine on a railroad, or the tread-power applied to drive a grindstone. The ordinary stationary steam-engine, or a saw-gate driven by gearing or by a belt, requires a fly-wheel of a size and weight proportioned to the momentum of the reciprocating movements. But a saw-gate or other reciprocating motion taken direct from the water-wheel by a crank on the same shaft will run well enough with no other fly-wheel than the balance. In fact, gates are often run in this way without either fly or balance; but a balance should always be attached when circumstances admit of it, because in addition to the assistance rendered by balancing the weight of the gate and pitman, another advantage is gained in the reduced strain and wear upon the binder and plumb-block, by which the crank is much easier kept in its bearings. The rule among millwrights for the weight of the balance is to have its balancing power as near to that of the gate and pitmen as possible; the extra weight of the saws, stirrup-irons, and gauges being found sufficient, with the advantage of gravity, to compensate the cut of the saws.

In computing this balancing power regard must be had to the distance of *each* from the centre; that is, the weight of gate, &c., must be calculated by the length of the crank from the centre, while the weight of the balance must be calculated by the distance from its *centre* of *gravity* to the same centre, which is generally greater than the length of the crank. To make this plainer, suppose the crank to be twelve inches long, and the centre of the balance eighteen inches from the centre of the shaft, then every pound weight in the balance will be equal to one and a half on the crank.

For a light and swift crank motion, such as a screen or sieve in a grist-mill, or the whiffers which take the rolls from a wool carding machine at the doffer cylinder, springs are an excellent compensating medium.

There are many circumstances in which a fly-wheel may be advantageously applied in a strictly rotary machine, either when the propelling power is unsteady or disturbed by other machines, or when the power is steady and the work or resistance is unequal, as in circular saws used for cutting firewood, making shingles and staves, or other work where the saw is alternately cutting and running empty at short intervals. By lengthening the arbor of such saws so as to place a fly-wheel upon the end at a convenient distance, and out of the way, a great improvement may be made in their working, as it tends to equalize the velocity while the saw is running idle, and giving it out again while the saw is cutting. This effects a considerable saving of power which is stored up as it were in a reservoir and given out when required, thus enabling a slight motive power and light belt to carry a saw through a cut, which without the fly-wheel would check up the saw or slip the belt, and be ticklish and troublesome to feed.

It should be remembered here that the principle of the fly-wheel is sometimes misunderstood and misapplied as well as that of the lever, and that the fly-wheel can never, under any circumstances, add power, but only equalize it. A machine may, therefore, be made with *too much* fly-wheel. An instance of this will show best what we mean. A friend of ours took a fancy for turning, and employed a good machinist to construct a crank power to be worked by hand to drive his lathe. The workman first set up a large shaft with a crank on one end and a fly-wheel on the other; the fly was heavy enough for a ten horse power engine, and a train of cog-wheels *ingeniously* contrived to gain speed by an advantageous leverage without losing power, connected this first shaft with the cone shaft from which the lathe belt got its motion; the result was, that three men sweating on the crank gave the first shaft a motion like the shaft of an overshot wheel, and drove the lathe like a buzz. But the men complained that the work was too heavy, and we were consulted to see if the work could be lightened; the result was, that the great *generator* of power, the fly-wheel, with its complicated train of cogged levers was set aside and a light band wheel upon the crank shaft substituted, from which the cone shaft was driven direct, and one man drove the machine with perfect ease and regularity.

Now this blunder was not made by an inferior mechanic, as the workmanship of the various parts was excellent, but was the result of mistaken theories most likely derived from a careless perusal of books and *jumping* at conclusions, which he had never enjoyed the opportunity of rectifying by experience. It may further be remarked here, as a general rule, that when a fly-wheel is necessary to any revolving machine it should be either upon or as near to the last and quickest mover as possible, and never, as in the case referred to, upon the first and slowest, where its effect is only to load and lumber the machine, and increase the friction, without any compensating advantage.

We will end this subject by a remark which we forgot to make when treating upon the saw-mill crank balance, which is, that the weight of the balance should never be *more* than the proportion there indicated, because the balance, although counteracted by the weight of gate and pitman, when up or down, has no compensating equivalent while acting horizontally, except the butt end of the pitman, which leaves a great centrifugal force unbalanced, and acting alternately in both directions at each revolution, has an injurious effect upon the binder and bearings; and further, that the balance and crank should be connected, and opposite each to the other, and not on separate parts of the shaft. A gang shaft was broken where we were working last winter, when no other cause could be assigned than that the balance was placed upon the tail end of the shaft, which was thus, in addition to the strain of driving, made the medium of connection between the crank and balance, and it snapped off at the crank bearing.

Centrifugal Force and Circular Motion.

Anything revolving around a centre has a tendency to fly off from that centre and proceed in a straight line; and this principle of motion is called centrifugal force. If we whirl a sling around rapidly, with a stone in it, and allow the stone to escape, it flies off in a straight line until its velocity is overcome by the resistance of the air and the attraction of the earth. A pail of water may be swung up and around in a circle without spilling the water even when the bottom is upward; or by turning the pail rapidly in one direction, the water is piled up against the edges of the pail, leaving a deep hollow in the middle. When grain is dropped into the eye of a revolving millstone, the motion of the latter gradually communicates motion to the grain, and it is impelled around and outward toward the edge of the stone, where it escapes as meal. It is this force that throws the water from a revolving grindstone, and sometimes, when the velocity is increased beyond the strength of cohesion in the stone, it is thrown asunder, and the fragments fly off like stones from a sling.

This tendency to fly off in a straight line from a motion around a centre is called centrifugal force, as already stated; but the power which holds bodies to the centre, and prevents them from thus flying off in a tangent to the circle they are describing, is called the centripetal force. And all bodies moving in circles are continually acted upon by these opposing forces; the ratio of each being alike and governed by the following laws—that is, these forces are always proportionate to the weight and velocity of the revolving body: If the weight of material be increased, its distance from the centre and velocity remaining the same, its centrifugal force will be increased in like proportion; if the distance from the centre be increased, while the weight and time of revolution remain the same, these forces will also be increased in the same proportion. If the number of revolutions in the same time be doubled, the distance from the centre and weight being the same, the centrifugal force will be four times as great; if the number of revolutions be increased three times, the force will be nine times as great, if four times, it will be sixteen times as great; and so it continues to increase as the square of the number of revolutions. This shows the rapid increase of centrifugal force with the increase of motion, and the necessity of lightening, strengthening, and balancing, all rapidly revolving machinery. It also shows the reason why a revolving machine, as a millstone, may be perfectly balanced while at rest, and yet be far from balanced when under a rapid motion. This is explained in the article on balancing millstones.

The centrifugal force is sometimes an auxiliary power in some kinds of reaction water wheels. This is the case when the head or pressure of the water is very great; the quantity and weight of water used for a given power being then small. When the head of water is low, the quantity required to produce any considerable power is so great that its *vis-inertia* altogether overbalances the slight increase of pressure induced by the centrifugal force. In other words, it requires a greater *power* to take such a continual mass of water as this requires, from a state of rest, and impart to it the requisite revolving motion while passing through the wheel, than all the auxiliary power yielded by the centrifugal force gives back. This fact we have established by varied and often repeated experiments; not with models, but with working machines, and will not attempt to explain it here theoretically.

The centrifugal pump illustrates the effect of this force upon water. It is a hollow tube, set upright in the water to be raised; two hollow arms are extended in opposite directions, from the top of the upright their hollows communicating, and the outer ends of the arms bent down so as to discharge the water into a circular trough placed under them to receive it. The upright tube has a valve, opening upwards, in its lower end, to retain the water when the machine is not in motion; it has also a journal in each end, upon which to turn, and a sheave, or pinion on the top, by which it is driven. To start the machine, the discharging openings in the ends of the arms are slightly closed, and the interior of the tubes filled with water through an opening at the top (centre); the opening is then closed, and a rapid rotary motion is imparted to the machine, which drives out the temporary plugs from the ends of the arms, and a constant discharge of water is thrown as long as the motion and supply of water are continued.

It is this centrifugal force that gathers and communicates motion to the air in every description of blower, where the blast is raised by a revolving fan; and a slight modification of the fan blower makes it a convenient and effective machine for raising or giving motion to water both by suction and force.

The centrifugal gun furnishes a better illustration of this force than any yet given. A carriage wheel whirling rapidly along a muddy road throws the mud from its circumference with considerable force. Boys will take a ball of tough clay and fasten it on the end of a small stick, or a small apple or potato is better than the clay, or a small stone stuck in a cleft in the end of the stick; by swinging the stick rapidly, the missile is thrown from its end with a force proportionate to the velocity and length of the stick. This is the principle by which the centrifugal gun throws its bullets; instead of a stick a hollow tube like a gunbarrel is used, one end secured to the centre of an upright shaft, to which a very rapid rotary motion is given by machinery; the other end can be raised or depressed, to increase or diminish the elevation of its rangè; the bullets are let into the tube at the centre, and thrown out of the other end with a velocity and force proportioned to the number of revolutions, and length of the barrel. This would be a tremendous engine of destruction if it could be manipulated so as to direct its balls to any particular point, as by fixing a reservoir of balls at the centre with a means for regulating their admission into the tube, a rapid and continuous discharge can be kept up, and by increasing or diminishing the velocity, and elevating or depressing the outer end of the barrel, the force and distance of range can easily be controlled.

When we were engaged in employments that required frequent correct estimates of the central force to be made, we fixed a unit or starting point, to compute from. We took sixty revolutions per minute, or one in a second, for the velocity, and found the distance from the centre at which that velocity would equal gravitation: but have now forgotten whether the circle described was ten inches in diameter, or a ten inch radius; we will therefore describe the simple expedient by which we determined this point.

Take a thin piece of board of any convenient width, and bevel one end to the angle of 45°, that is, to a mitre; slip this through a slit in a vertical shaft, with the short corner of the mitre up; now fasten a weight by a string or small wire to the upper short corner of the mitred end, and set the machine in motion. Time and regulate the revolutions to 60 per minute, and then knock the board backward or forward in the slit, to lengthen or shorten the carrying arm, until the string supporting the weight corresponds exactly with the bevelled end of the arm. This shows the centrifugal force to be equal to the weight, in other words, it is drawn *outward* by centrifugal force, just as much as it is drawn *downward* by gravity: and it only remains to measure the distance from the centre of motion to the centre of the weight (not its point of suspension) to complete the problem. This velocity and distance, if retained, give a basis from which to calculate for any other velocity and distance from the centre, by the rules of ratio and progression already given. The *weight* of the mass under consideration, whatever it may be, furnishes, of course, the other important item in all such computations.

Action and Reaction.

These forces are always equal, and in contrary and opposite directions one to the other, whatever the power or force employed to generate the motion may be. The powder exploded within the gunbarrel or cannon exerts the same force against the breech as against the bullet or missile shot from the bore; the difference in velocity between the projectile and the recoil of the gun being as the difference of weight in each. The recoil or rebound of the guns in a ship of war has frequently kept them off a lee shore when all other resources had failed, and has sometimes set them afloat when they were actually aground. The chute of water projected against a wheel exerts the same force against the penstock or flume that it communicates to the bucket upon which it impinges. This is the power employed in Dr. Barker's and other reaction wheels.

A horse in the act of drawing exerts the same force, in a contrary direction, against the road that he does in moving forward the load. This is the power employed in all horse tread-powers, whether the animal be hitched to a bar, and draw on a horizontal wheel, or climb an incline.

A better illustration of this law of motion will be found in rowing or pushing a small boat upon the water. The oar being pushed against the water moves the boat in one direction, the "slip" of the oar moves the water in the opposite direction; the force and motion being divided and absorbed somewhere between these—one representing action, the other reaction. Now, suppose the oar be placed against earth, or something immovable, instead of the water, and the same force applied: the pressure, as before, will be equal and opposite, but all the motion is given to the boat—the result being the same whether the man be in the boat or on the land. To make this plainer, suppose two boats of equal size and weight, with a man in each; connect these by a long light pole, each man having hold of the opposite end. Now let both men pull alike upon the pole, and the boats will approach each other with equal speed and force, meeting half way; if something elastic, as a suitable spring, were interposed between their points of contact, it would be compressed until the forces of both were overcome, when the elastic spring would react, sending both boats back apart again; or rather let both men cease pulling when the boats approached sufficiently near each other, and push apart; they must now exert the same force to stop the motion that it took to engender it, excepting the loss by friction; and if that force is continued after the approaching motion is stopped, then the boats will begin to move apart again, as in the other case; the strength of the men exerted through the pole being interposed instead of the spring. In this case, the boats being of equal size and weight, the motion of both will be alike. If the size and weight of the two were very different (as in the case of the gun and missile shot from it), then the velocity of each would be as the mass and weight of each.

The small steamer "Nellie Tupper," four horse power, when towing logs on Chateaugay Lakes, left the two men, who made fast the tow-line to a raft, upon it; the captain refusing to stop the boat, or send a small boat to take them on board. Both being strong resolute men, they fixed themselves in an advantageous position, and undertook to pull the steamer *back*, and get on board; this they accomplished to their own satisfaction and of all who witnessed the feat. But all they actually accomplished was to bring the raft and boat together by pulling on the tow-line, like our supposed men pulling on their connecting pole. Here note particularly this difference, which is of vital importance to the millwright; the two men pulling on *one end* of the tow-line, had a *power* equal to four men, two pulling at each end, but only half the *speed*. Here the boat and raft both continued their journey, the speed of the raft being accelerated, while that of the boat was retarded, in exact proportion to the mass of each.

We have neither the ability nor wish to write a philosophical treatise upon Action and Reaction, but as it is a subject that every millwright should understand perfectly, we will try to make those points a little plainer, which particularly relate to his business.

If two boats, or equal other floats, be placed side by side upon the water, and a man stand upon one and push lengthwise upon the other, if both be alike free to move, the one under his feet will move back and the other forward as he travels along, each moving at half the rate of the man's walk. If the one upon which he stands be fastened, the other will be pushed forward at the same speed the man travels. If the one against which he pushes be fastened, and the other released, then the one upon which he travels will move at the full speed of his walk. In the first instance each float moves at half speed, but in opposite directions; in each of the other cases only one float moves, but with the whole velocity of the man's walk, the power applied in the three cases being the same. But let the man push against some part of the float on which he stands ever so hard, and he can give it no motion. This would be like working a great bellows placed on the after end of a vessel to fill a sail on the bow, expecting to drive the vessel forward—such an arrangement could give it no motion, but if the sail were removed, and the bellows worked, the vessel would be driven backward. The wind from such a bellows or other source, clear of the vessel, impinging against the sail, would of course give it headway.

Now many ingenious millwrights have attempted, and some claim that they have actually constructed, a water wheel that will combine both of these powers, and give a double result. We confess to pursuing that "ignis fatuus" in our own younger days, but we never succeeded in combining both powers in the same wheel, which would be like getting into a basket, and attempting to lift one's self by the handles, implying a genius of the perpetual motion calibre, which we now disclaim. We placed a light tub wheel outside of Dr. Barker's reaction tube wheel, which intercepted the discharging water, and of course revolved in the contrary direction, but found it much better to lengthen the tubes or arms of the first wheel and take all the power from that, than to divide the power in two, and gear those two separate powers together upon the machinery to be driven, as it made a very good and simple arrangement very complicated, and detracted very sensibly from the power. See the article on Barker's wheel.

There is a way of employing the recoil or reaction of the water upon the same wheel, after or *as* it *strikes*, which makes the force of impingement similar to that produced by the stroke of a perfectly elastic body. To effect this, the water must be sent around a curve representing a half circle, the water being shot into one end of the curve, which it follows around and escapes from the other end with nearly the same velocity with which it entered, returning toward the point from which it issued, like the curved plate armor which Louis Napoleon placed upon some of his ships of war, to conduct the balls fired against them around a similar curve, and hurl them back in the teeth of those who fired them.

To test the actual gain by this plan, we caused a column of water to descend perpendicularly upon a level scale, and balanced its force with weights; we then put a curved scale, such as indicated, in place of the flat one; this returned the column of water up again, close to the descending column, but not touching it. It now required double the weight to balance the scale. Some would think this a very valuable discovery, but this double power, which seems so sure and satisfactory when weighed and tested upon the stationary scale, is very much modified and dissipated when practically applied, and the motion of the wheel as it recedes before the water, carries a great part of the recoil away with it, and changes the exact half circle into the curve seen in the buckets of our centre vent water-wheel, which see in the chapter on Centre Discharge Wheels.

The proper curve required to catch and return the water so as to get the benefit of this recoil, is very difficult to understand, and still more difficult to explain. The water is affected by so many different contingencies and circumstances in passing through the curved bucket while the wheel is in motion, that we only found the proper curve by many repeated experiments, as detailed in the description of the wheel. In the first place the water must be shot into the outer end of the bucket, that is, at the circumference, and not at the centre of the wheel: because then the water passes through in the direction of the wheel's motion, the friction thus helping, instead of hindering, the motion of the wheel; and also because the centrifugal force acting against the force of the entering water, forces it outward against the inner *driving* side of the bucket, and prevents it from touching the back of the next one; and lastly, it brings the discharging water near the centre of the wheel, where the motion is comparatively slow, and this allows the nearly spent water to impinge its little remaining force upon the short curve of the inner end of the buckets, by its recoil.

In referring to the recoil of a gun, we said it was equal to the action of the powder against the ball—that is, the velocity of the ball multiplied by its weight is exactly equal to the velocity (recoil) of the gun and its attachments multiplied by their weight. This may seem to be following the laws of weight and velocity to an extreme; but it is sometimes necessary for the millwright to deal with this law in a still greater extreme, like firing the same charges of powder without any ball, or with blank cartridges. In this case, the powder having nothing but the atmosphere to act against, the recoil is very small; give it water instead of air to act against, and it will shiver the hardest rocks, although otherwise unconfined. Large rocks in the rapids of the river St. Lawrence have been blasted and removed in this manner, in situations where it was impossible to drill and blast by the ordinary method. Many years ago, we saw a rotary steam engine driven by the recoil of the steam, on the same principle as Barker's water wheel. A pipe from the boiler carried the steam up into a central hub or chamber, from which two hollow iron arms projected in opposite directions; the hollow of these communicated with the interior of the hub and admitted the steam, which was forced out through a small hole in the opposite side of each end. These arms were about three feet long, flattened to avoid the resistance of the atmosphere, and slightly bent at each end. They resembled, in shape and size, the iron scabbard of a dragoon's sword, and the hole in each for the discharge of steam, the size of a large knitting needle. The outside of the hub was turned off to the shape of a pulley, and a belt from it drove the machinery. This is the cheapest, most direct and simple of all the applications of steam to drive machinery; and if steam were a dense and heavy fluid, such as water, it would also be the most effective. But this is the difficulty, steam is so light and ethereal, having such facility for increased motion and escape under high pressure, as shown by the action of the steam-gun, that this mode of working it is like exploding gunpowder in a blank cartridge. This engine made somewhere about fifteen thousand revolutions per minute, and used all the steam that could be raised in a boiler rated at twelve horse power.

The steam gun referred to threw a continuous stream of bullets from a hopper at the breech, and it was found that every increase in the length of the barrel used made a corresponding increase in the distance of its range, which is not the case where gunpowder is used. Perkins, the inventor, claimed that it could throw a ball from Dover, in England, across the channel, to Calais, in France, a distance of twenty two miles.

The centrifugal gun is another engine that throws a continuous stream of bullets from a hopper at the centre. It is simply a long tube like a gunbarrel, one end being fastened to a central pivot or shaft, which communicates to it a very rapid revolving motion. The balls are admitted into the tube at the central end, and thrown out at the outer revolving end by the centrifugal force. The principle is similar to that of the sling.

It occured to us that a combination of these two machines could be made, by using two round tubes like these gunbarrels, as arms for a rotary engine, which would yield a very strong reaction power. To do this the tubes would have to be bent by an easy curve towards the outer end, to a right angle, so as to discharge on a tangent with the circle described, and we mention it to illustrate the principle and mode of reasoning which led us to adopt the following modification of the reaction rotary steam engine, which makes it a very effective and economical machine.

The machine itself, that is the central hub and hollow arms are longer than that described before, as it requires a greater interior capacity, but otherwise it is the same, and the modification consists in forcing water through it instead of steam or bullets. The weight and incompressibility of water enable it to resist every acceleration of motion. A double pressure of steam would double its velocity and escape, and only double its power of action and reaction; but it requires a fourfold pressure to double the velocity and escape of water, and this double discharge of water gives *eight times* the power of action and reaction. For an explanation of this, see the article on the "Pressure and Power of Water."

The pressure of the steam is made to act upon the water on the principle of Savery's double engine. See the chapter on "The Transmission and Transportation of Motive Power."

So long as the mechanical power of water was obtained almost exclusively by the use of the overshot and undershot wheel and their modifications, a millwright might get along pretty well without a very thorough knowledge of the peculiarities of this action and reaction of water, and the relation of each to the other; but now, that such a great variety of wheels are used, employing every modification of these two forces, and acting upon principles more or less scientific and intricate, a more perfect understanding of these principles is indispensable.

There is one universal law to be accepted in every application of motive power, whatever that power may be, which is: That the motor, being limited both in force and velocity, may be made to give out nearly its whole force in one direction, and its whole velocity in the opposite, like the gun, and missile shot from it; or it may be made to yield a certain force and velocity in both (opposite) directions, like a rocket, ascending by the combustion of its contents, reacting against the air. But whatever force and velocity it yields in one direction are subtracted from that in the opposite direction, because each velocity and force, being measured and multiplied together, and the product of both added, give the measure of the motor applied, and it is not possible to make any combination, by which more than this measure can be obtained.

For an illustration of this, we will suppose a Barker's wheel, Fig. 25, discharging its water upon the buckets of another wheel placed around its outside, and both geared to the same machinery. The outside wheel receives the direct action of the water, the inside one the reaction or recoil. Here it is evident that the velocity with which the inside wheel moves must be deducted from the velocity of the discharging water before it strikes the outer wheel; and further, that the proportions of the gearing connecting these may be changed, so as to give one nearly all the velocity, and the other nearly all the power, and *vice versa;* and that either will yield all the power and speed of both by stopping the other.

Friction.

The rubbing of one surface against another is called friction. It has a constant tendency to diminish power, and retard velocity in every description of machinery; and in adapting the power of any first mover to a particular kind and amount of work, an allowance of more or less power, according as the machinery is complex or simple, must be made to compensate for the waste of power and velocity by the friction of the moving parts.

There are so many contingencies affecting the different kinds of machinery, and the conditions and circumstances under which they are worked, that no rule that would be even approximate, can be given to compute the amount of waste by friction; some machines, of simple construction and good materials and workmanship, using up $\frac{1}{20}$ part of the motive power; while others, where these conditions are all the opposite, use up one-half or more, of the power applied. The friction of polished wood or metals sliding on each other, is equal to about one-fourth of the pressure. The friction of wood and metal working upon each other, is a little less than that of two woods or metals working in contact. Friction is considerably reduced by interposing some kinds of unguent, as tallow, oil, or water, between the rubbing surfaces. Tallow is found best for wood, as oil or water tends to open the pores or grains of wood, and soften its texture. Tallow is too hard for iron journals, unless in some cases, as in steam-engines, where the heat is sufficient to keep the tallow fluid. Olive and sperm oil are both good for lubricating all kinds of metallic journals and bearings. Hog's lard is a good substitute for oil between metals, or for tallow between woods, or wood and metal. Powdered black lead mixed with the lard, tends to polish and harden the wood by filling the pores.

Water, when clean, and never allowed to get dry, is the most reliable lubricator for the iron journals of water-wheels, when run upon wooden bearings; but ample provision must be made to admit the water between the journal and step or bearing, or they will heat, and burn out the step in a very short time, even when deep in back-water.

The retarding effect of friction is not materially increased or diminished by either enlarging or diminishing the surfaces of contact; but by lengthening the bearing of journals, or other movements, the friction is distributed over a greater surface, and they will keep cooler, the oil will keep better, and the danger of grinding or wearing the surfaces is very much lessened.

Neither is friction affected in any material degree by changes of velocity, its retarding effect being nearly the same at all degrees of velocity. But to increase the velocity by enlarging the diameter of a journal, or diminish velocity by reducing its (the journal's) diameter, will increase or diminish friction in the same proportion, because the increased diameter moves the resistance of the friction further from the centre, and gives it more purchase to hold back, while the diminished diameter moves it nearer to the centre, and thus diminishes its power of resistance, while the friction at the point of contact remains in each case the same.

This will be better understood when we consider that the power of friction to resist the motion of a revolving wheel is directly as the diameter of the journals and inversely as the diameter of the wheel.

Although friction is not materially affected by changes of velocity, or by altering the breadth or area of bearing, it is very sensitive to any increase or diminution of weight or pressure; and any variation of this in either direction increases or diminishes the friction in the same proportion. This shows the economy of making machinery as light as is compatible with sufficient strength.

Friction, while its constant tendency is to obstruct and destroy motion, may, nevertheless, be sometimes turned to useful purposes. It is friction that enables a wedge to hold its lift, as each successive stroke advances it to the head. It also enables the screw to hold each advance made by a turn of its lever. Friction also furnishes a convenient medium of communicating and transmitting motion in machinery, as in gigging back the carriage and log in saw-mills; and in some modern mills, the whole driving power for both saws and mill-stones is communicated by friction of iron upon iron. It is friction also that enables belts and chains to convey the motion from one pulley to another.

5

CHAPTER III.

TRANSMISSION AND TRANSPORTATION OF MOTIVE POWER.

It is frequently necessary to transmit motive power to a distance from the source where it is generated. This is generally effected by means of shafting coupled together, or otherwise connected, or by belts or chains carried over pulleys. Sometimes it is transmitted through connecting rods of iron or wood. We have seen it transmitted for several hundred feet from one iron shaft to another lying parallel to it, each shaft having three equal cranks, exactly alike, and connected by iron rods in such a position that one of these cranks was always in a position for the connecting rods to *draw* upon the crank at its opposite end; and thus, by the combined action of the three, the motion of the first shaft was communicated to the second. This plan was used to transmit the power from a water-wheel in the rapids of the Niagara River to a grist-mill upon the top of the rock. This mill stands (or stood) upon the American side, just above and in sight of the suspension bridge. In all these cases the power is transmitted through the various mediums *after* the motion is generated, and the distance to which it can be carried by any of these is very limited, as the weight of material and the friction soon use up the motive power. Probably the greatest distance that motive power was ever carried was by a combination of jointed rods, similar to that at Niagara. These were used to connect a series of pumps with the water-wheels which drove them, at the celebrated water-works of Marli, near Paris, in France. Eighty-two of these pumps were placed more than three hundred feet *above* the power which drove them, and *half a mile away*. The sound of these rods working was like that of a number of teams loaded with bars of iron running down hill, with axles never greased, and it was estimated that 95 per cent. of the power was wasted in communicating motion to the machinery.

There are some motors that admit of transportation from the place where they are generated, and then transmit their power to the machinery in the situation in which it is to be used. These are water, steam, and, to a limited extent as yet, compressed air. The situation in which water powers are most generally found being in low ravines, precludes its transportation through tubes or conduits to the places where the power could be most advantageously used, except in very rare cases. From the nature of the fluid, the place to which it is carried must be still lower, or at all events no higher than the place where the power is found. And then it is so dense and so heavy, and resists any great acceleration of motion so effectually, that large and strong tubes are necessary to conduct it in quantity sufficient to yield any great amount of motive power. And the expense of these tubes, if nothing else should interfere, would set a limit to the distance to which water could be carried for the mere purpose of applying its motive power to propelling machinery.

The transportation of steam, in a like manner, is objectionable, although the objections against this are different. The elasticity of steam, and its great facility of motion when compressed, admit of a great amount of power being transmitted through a comparatively small tube, and this in any direction, without distinction, up or down. But the trouble in conducting steam in this manner is the constant tendency to waste by condensation; the effect of the absorption of heat by the pipes and their surroundings. This can be partially counteracted by inclosing the pipes with felt or some other non-conducting covering; still, the waste from this source alone is so great that it prohibits the conduction of steam power in this manner, except through very limited distances.

The conduction of motive power through tubes by *compressed air* has been often proposed, but this has been practically tested only to a very limited extent, and generally on a small scale. That this should be the case is rather singular, when we reflect that this medium has all the facilities for rapid transit through tubes that steam possesses, and nearly in the same perfection; while, on the other hand, it is free from the objections inseparable from the transportation of either steam or water. It is not subject to waste by condensation, like steam, neither is it affected by frost, like water.

The protection of such conduit pipes for water, as we are considering against frost, in all cold climates, would nearly equal the cost of their construction; and without such protection they would be rendered inoperative in cold weather, and the tubes ruptured and ruined.

Air is not affected to any material extent by the ordinary changes of temperature, and the principal reason why it has not been more used in a condensed state, for the transmission of motive power, is the mechanical difficulty of confining and working it in the condensing machinery and that by which the power is given out at the place to which it is transported. The reason is, that it is a fluid so ethereal, and possessing such a facility for rapid escape through a small crevice, when strongly condensed, that it requires the working machinery to be made and maintained in the most perfect manner, to confine and compress it economically. Yet this tendency to rapid transit through a small issue, which requires such care in the adjustment and packing of the machinery to work it, is the very property that renders it so valuable a medium for the transmission of motive force; and the aspect of this whole problem is changed when we consider the facility with which the pressure of *water* is transferred to *air*, and that of the *air* transferred back again to *water*. The combination this suggests, of compressing air through the medium of water, where the power occurs, and transmitting the pressure back again to water, where the power is to be used, employing the air only as the medium of transmission, dissipates the difficulties which pertain to either fluid when used singly, and utilizes the valuable properties of both when thus combined.

We were brought to adopt the combination indicated, as the best means of transmitting water power, not by philosophical research, but by an accidental train of circumstances; and as a detailed account of these will help to illustrate the subject, we shall try to give it. We owned a mill, where the quantity of water was small, but the fall was high; and we spared neither time nor money in our efforts to utilize the whole power. When a very dry time occurred, the power was still insufficient for the work and had to be supplemented by steam, an ordinary ten-horse stationary engine. This engine, more particularly the boiler and furnace, was not of good construction, and yielded a poor return for the wood and attendance; this result we ascribed to the disadvantageous application of the steam through the cylinder and piston, the crank and fly-wheel; the latter appendage being exceedingly heavy and cumbersome, and we set to work to invent a steam engine that would produce a rotatory motion direct from the steam, and dispense with all these and their unavoidable friction. We were aware of the defects of the reaction rotary mentioned in the chapter on "Reaction," for we had seen that tried, but were not aware that any one had ever made one to act by *confined* steam, on the principle we meditated. The result was, that we succeeded in perfecting the engine (a simpler one than any we have since seen or heard of), but were terribly chagrined when on looking up the means and forms to secure a patent for the invention, to find that it was an old device that had been repeatedly tried, and found comparatively worthless.

This disappointment turned our experiments in another direction. We had previous to this succeeded in applying the force of water under a high pressure economically, and were acquainted, by the various experiments made, with the rapid increase of power which a small quantity of water could be made to yield, by increasing the pressure to which it was subjected. And we concluded to turn the full pressure of steam from a boiler direct upon the surface of water confined in a strong tank, and apply the water as it was forced through a pipe from the bottom of the tank, to propel the machinery. After our plan was matured, and arrangements made for carrying out the design, subsequent business affairs intervened, so that we did not proceed with the alterations; but we have never had any doubt that the arrangement can be carried out to work satisfactorily and give a good result.

The following description and plan will give an idea

of the machinery intended to be used, as shown in Fig. 21. The tank is double, that is, twice as long as it is wide, and divided through the middle by a partition. A

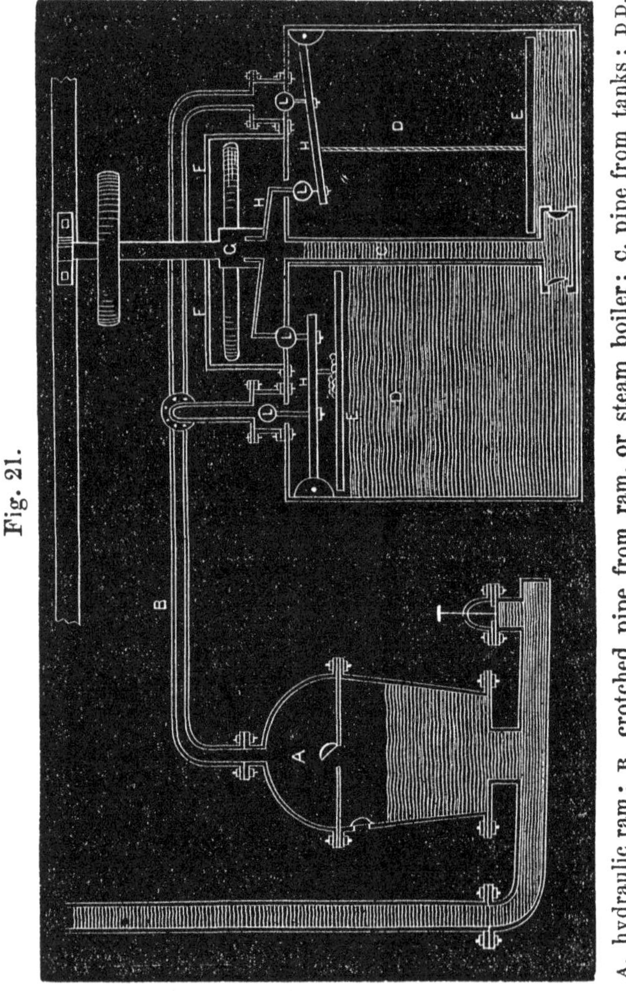

Fig. 21.

A, hydraulic ram; B, crotched pipe from ram, or steam boiler; C, pipe from tanks; D D, tanks; E E, followers; F, wheel-case; G, reaction engine; H H H, valve levers; I I I I, valves.

tube, of a bore sufficient to pass the intended quantity of water freely, is put up from the bottom, alongside of this partition, at the middle. This tube communicates with the two interiors at the bottom, by a valve in each side opening inwards, and its upper end passes through the top of the tanks. It will be seen by reviewing this arrangement, that a pressure inside of one tank would open the valve at that side, and close the valve at the further side of the upright connecting tube, and the contents of that tank would be forced up the centre tube. When one tank is emptied, and the pressure let in to the other, the play of the valves is reversed, and the contents of the second tank are forced up the centre tube, as before.

The motive power is obtained by placing a rotary reaction engine, like that described in the chapter on "Action and Reaction" in this work, upon the top of the upright centre tube, and forcing the water through that engine. A wide rim is placed on edge around and encircling the discharging arms, and covered to the centre shaft, thus forming a tight case to intercept the water, the top of the tank forming the bottom of the case. A valve opening downward in each side of the case allows the intercepted water to run down into the exhausted tank through its valve, while the valve in the top of the working tank is closed by the internal pressure.

The steam, or compressed air (it may be used with either), is admitted into each tank alternately, by a crotched pipe, a branch of which communicates with the interior of each tank, through an opening in the top, each opening being covered with a valve opening upward. These valves are attached by a short stem to the middle of the same lever, that controls the other two valves through which the water is returned from the wheel-case into the tanks. One end of these levers is attached to the inside corner of each tank by a joint, as shown in the figure; the other ends are connected by a small balance beam, to insure the two sets of valves reversing at the same instant. A loose floating follower is attached to the end of each valve lever by a rope of such a length that the weight of the follower in the emptied tank will assist the buoyancy of the follower in the full tank to shift the valves.

To set the machine in motion, one tank must be filled with water, the other empty to the top of the valve in the connecting tube; the steam is then let into the top of the full tank, its pressure closes the valve, which was already partially closed by the float, and, having no other means of escape, it presses down the float and forces the water up the tube and through the wheel. Here it is collected in the wheel-case and runs through the open valve into the empty tank; when the water is settled down to the top of the valve in the central tube, a rush of steam passes through with the water, filling the wheel-case and the space above the float in the other tank, which is now full of water. The steam escapes in this way so rapidly that the interior pressure is speedily reduced, and at the same instant turns the steam into the other tank, and the whole process is repeated from the opposite side. Thus the same water is alternately forced from one tank and collected in the other, and requires only a very slight addition to the quantity to supply the waste by evaporation.

This is a cheap and simple method of obtaining the motive power from steam. It is also economical, especially after the water has become heated by repeated use. The quantity of water wasted by evaporation increases with the heat, and as the water approaches the boiling point, the condensation in the wheel-case is not sufficiently rapid, and it requires an escape pipe in the top of the case for the exhaust steam.

Those who are familiar with the various applications of steam, will perceive that this machine works upon the same principle as Savery's double engine, which was invented and used for forcing water up out of the mines in England. A good description and history of this celebrated engine is given in "Ewbank's Hydraulics and Mechanics."

A number of years after studying out the plan we have given the details of for transmitting motive power by applying the pressure of steam, our attention was again drawn to the subject, but in the interval a new idea had occurred to our mind. There are numerous waterfalls in many parts of the country, one in particular, the "High Falls" on the Chateaugay River, quite near the village, which being the outlet of the Chateaugay Lakes, might be made to yield a power of several hundred horses. It occurs, however, in a deep ravine, which is almost inaccessible, and we were next led to consider the feasibility of erecting a powerful water wheel at those falls, and by means of a large force pump, driven by that wheel, to compress air into an air vessel, and convey the air thus compressed to the village, to supply the motive pressure for these engines instead of steam.

The rapidity with which compressed air traverses through a tube would enable one of a comparatively small diameter to transmit all the pressure required to drive the machinery we contemplated erecting at the time, but the *expense* of such a tube of suitable material, with that of the water-wheel and other condensing machinery, was beyond our limited resources, and so we banished the subject as visionary; and the reader would not now be troubled with this part of the subject, had not the "Scientific American" taken it up. But the fact of this means of transmitting and transporting motive power being sanctioned and indorsed by such an authority, gives us reason to hope that we may yet see this our favorite device brought into extensive operation.

In Vol. 19, No. 13 of the Scientific American, page 196, is given an abstract of an article from the *Bulletin Mensuel*, wherein M. Leloup discusses the subject in an interesting and comprehensive manner; also an engraving and description of the apparatus by which M. Leloup proposes to compress and transmit the air by the medium of water. It will be seen by a reference to that engraving and description, that he proposes to work the compressing machinery in contact with water, as we intended; thus obviating the necessity of any great degree of accuracy of workmanship or finish, except in turning the piston and fitting its packing box, and the necessary valves. The cylinder does not require to be bored, all that is required of it and the other containing parts, being sufficient capacity, strength, and tightness, like the tanks, &c. at the other end of the tube for giving out the power.

It is strange that a means so obvious of obtaining a cheap and reliable motive power should have been so long neglected, especially in localities where great water powers occur; and that two persons so differently situated should both be impressed with the utility of this device at the same time is a singular coincidence. One, a learned *savan*, exploring the regions of science and philosophy, and deducing new theories therefrom, the other an humble self-educated mechanic, seeking only to lessen the expense of driving his mill machinery, in order to reconcile revenue and expenditure. The only difference between us in tracing the problem to a solution appears to be, that the *savan* began at the fountain head and followed the current, while we began at the other end, and had to work up stream.

Another correspondent, signing "A." in Vol. 19, No. 14, *Scientific American*, suggests the use of the *trombe* or water bellows, as a means of compressing air for the generation and transmission of motive power. This machine only collects and compresses the air within the receiver by dragging a portion of the surrounding atmosphere along with a swift descending column of water, and is only available as a ventilator, or by urging the force of a draught to assist combustion, like a fan blower. It could not be made to yield more than one, or at the utmost, two per cent. of the power of the column of water by which it is generated, and is therefore not adapted to the purpose under consideration; but the same column of water, applied directly to compress the air within the chamber, instead of M. Leloup's piston, on the principle of the "*Hydraulic Ram*," would probably be the very best arrangement that could be made to effect the necessary compression of air, by the power of water. All the alteration which the ordinary ram for raising water would require to transform it into such an air compressing machine, would be to remove the ascending pipe from the water in the bottom of the condensing reservoir, to the air in the top of it, when the pipe, instead of elevating a certain amount of water at each stroke or pulsation, would transmit an equal quantity of compressed air, and a valve opening inward to admit a fresh supply of air at each respiration.

Such a ram would be preferable to M. Leloup's apparatus, for several reasons. It would dispense with the water-wheel, cylinder, and piston, and their connecting machinery; and this greater simplicity would lessen the expense, and also enable the same column of water to transmit a greater proportion of its power to the compressed air, than if applied through the medium of such intervening machinery; and last, though not least, it would dispense with the cost of a person to attend the machinery, because the water ram, as perfected by Montgolfier, and now so extensively used, is wholly self-acting, and continues to operate night and day, summer and winter, like the circulation in a living animal, until interrupted by some casualty or extraneous force.

We know the opinion is general that the principle of the water ram can only be applied on a small scale, but this is an error, caused by the fact that while small rams raising water for household purposes are so common, the principle is never seen applied on a large scale.

It is just as applicable on a large as on a small scale, the only limit in that direction being the strength of the materials of which the confining tubes and reservoir are composed.

CHAPTER IV.

PECULIARITIES AND PROPERTIES OF WATER.

The text-books on natural philosophy treat of water under the three different heads of hydrostatics, hydraulics, and hydro-dynamics. The first treats of water in a state of rest; the second of the principles and peculiarities of water in motion; and the last, being the branch which treats of the adaptation and application of water to machinery, is that with which the millwright is more particularly interested. Although this last division of the subject may be the most *interesting* to the millwright, it is important that the principles which govern the dead pressure of water, and its flow through confining channels under different circumstances, should be equally well understood. The books referred to are so common, and treat these subjects so full and familiarly, that it would be a needless waste of time to reproduce their explanations and illustrations here; and we shall only notice a few of the most important peculiarities of water under these different conditions.

Water, like every other fluid, has a constant tendency to rush toward the centre of the earth. This is the natural effect of the power of gravitation, which exerts a constant force upon all bodies, whether solid or fluid, according to their bulk and density. The cohesion of the particles composing a solid body enables such bodies to lay stationary and at rest in any situation where their centre of gravity or weight is supported; but fluids, not possessing such cohesion, and the particles of which they are composed being perfectly mobile and free to act independently of each other, each particle takes its own most direct route towards the centre of gravitation; and hence the necessity and difficulty of confining fluids within restricted limits, and their continual tendency to burst through or over such limits in any direction toward a lower level. It is this perfect mobility of the particles, and the independent freedom of each particle to obey its own impulse, that enable water to assume a perfectly level surface when confined by sufficient boundaries, and it is this peculiarity, acted upon by gravity, that gives it such velocity and impetus when it escapes through or over such boundary.

This also accounts for the facility which water (in common with other fluids) possesses of pressing with equal force in all directions, at any particular point or depth below the surface, and explains how it is that "a quantity of water, however small, can be made to balance another quantity however large." This is called in books "The hydrostatic paradox;" but it always appeared to us that the paradox was all in the form of the statement, and not in the water at all; because when water is thus situated, as in two tanks, one large and the other small in diameter, both interiors being connected by a tube at the bottom, any one can understand at once how it is that the water in the large tank stands at the same level as the water in the small one, or how the water in the spout of a teapot or kettle stands at the same level as that within the vessel itself, and it appears paradoxical, or rather impossible, that it should be otherwise.

There are several peculiarities and conditions pertaining to water which are never mentioned in books, but, as experience has taught us that some of these are both useful and interesting to millwrights, we will mention them here.

Taking Levels.

And first, the facility and accuracy of taking levels from the surface of still water. We have often known men who were contemplating the building of a mill, or mill-dam, or canal, to go many miles to get a millwright to level the site, or to lend them a spirit level, and give them some instruction to enable them to level and determine the amount of head that could be had, or the height of dam, bank, or building. Now such levels can be taken from the *surface* of *still water* more accurately than by any spirit level, and by the following process: Take two poles of sufficient length to reach from the bottom of the water to the height of the required line of level, measure these poles or laths from the upper end down to the length intended to stand *above* the water, and make a plain notch or mark upon both sticks at this point, by laying both together, to insure perfect equality of height. These may be marked in feet and inches, for convenience in showing, or in varying the line of level. Now point or sharpen the lower ends of the poles, and stick them down through the water into the earth at the bottom, until the notch marks are both at the level surface of the water, taking care to have them stand plumb, and in the right lines, and at a convenient distance apart; then sight across the top of these two and set a third, and fourth, or any number required to run the line of level to the desired point, ranging the tops accurately by the first two; and the tops of these poles will show a water level, so many feet above that of the water from which it is taken.

If the poles are all measured from the top end, and marked, each one will show at a glance the relative height of the ground on which it stands, whether above or below that of the water. Another advantage in having them measured and marked, is that when running the line of level down stream, it can be dropped (lowered) to any of the marks below, whenever the height of poles becomes inconvenient, the number of feet dropped each time being noted, and counted in the final result. It will be easily seen that when the line becomes inconveniently low, as in running it up stream, it may be raised in the same way, the amount of rise above the original line being accounted for. It is obvious, that by shifting the position of one or both of these poles in the water, the line of level may be run in any other direction, and equally so, that, as the poles from which the level is taken may be any number of feet apart, or as many rods, not exceeding the range of accurate vision, a more exact level is obtained in this way than can be taken by an ordinary spirit level.

Fitting down Sills under Water.

It frequently happens that sills have to be fitted down to the bottom of the water, for the erection of mills, flumes, dams, bridges, and other structures, in situations where the water cannot be turned aside, or kept out without great expense for coffer-dams, pumping, &c. The following method of obtaining an exact outline or profile of the bottom in all such situations, we have practised for many years with invariable success:—

The first requisite is a level surface on the water under which the sill is to be placed. To obtain this it is sometimes necessary to obstruct the surface current sufficiently to back and deaden the water as far as the sill extends; now fasten up a row of stakes along the intended bed of the sill, and nail a wide, thin board upon edge to these stakes, the entire length, the lower edge at the surface of the water (water line). An exact outline of the bottom, or bed of the sill, is transferred to, and marked upon the board by the following process: Fix two pieces of wood in the form of a T square, the tongue-piece longer than the depth of water, and marked with feet and inches, like a ten-foot pole, the T head about two feet long, and three or four inches wide, with a mortise through the middle in which the tongue-piece can slide freely up or down, and at the same time be kept plumb. Place the T head on the edge of the board, and slip the tongue-piece down through the mortise to the bottom, and try the depth along the whole bed, until the deepest spot is found. Here cut a notch or make a hole in the tongue at the surface of the water, to hold a pencil, and let one man hold the T head with one hand, and the tongue with the other, moving both hands carefully along the board towards one end, feeling the bottom as he advances, while another man holds the pencil in the hole, marking all the rises and inequalities of the bottom upon the board, taking care to mark only when the tongue-piece touches the bottom. When thus marked to one end, commence again at the same low place, and mark to the other end; and the outline of the bottom will be transferred to the board, with the relative level of either end or any other part indicated by the distance of the pencil mark from the lower edge of the board, which is the water line or true level.

The marked boards may now be taken down, and the portion above the pencil mark cut away, when the other portion left will be a pattern by which to fit the under side of a sill to the rock or bottom. It is sometimes better to take the pencil line some inches above the water line at the lowest part, and where the bottom is very uneven it is best to mark the sill by the pattern, so that only the slight inequalities will be cut out of the stick, which will not affect its strength, and where low spots in the bottom occur, short pieces should be spiked on to fill out the pattern. By this plan, such timbers can be fitted to a rock, or other bottom, under water, nearly as accurately as if dry, and with very little more expense.

This method of scribing down a mud-sill under water, and the method of taking levels from water, as described in this chapter, are, as far as we are aware, our own inventions, but the following plan for

Washing out a Mill-race or Foundation by Sluicing

we learned from a friend, a returned Californian. We had seen several attempts made to wash out a race or a foundation by water, and had once tried it, but all these attempts ended in confusion and vexation, for the reason that they were all begun at the wrong end. That is, the water was shot down from a height above into the foundation, or upper end of the race, without any proper facility being provided for carrying away the gravel and small stones; the result was always the same, the water would excavate a deep hole where it struck, but only the soft loam and clay were melted and carried away by the water, the stones and sand remaining and blocking up the channel, and it was found cheaper and better to plough and scrape the earth out. But Mr. Whipple, the Californian alluded to, has shown us the principle of *sluicing*, which experience has proved to be the cheapest and best method of making such excavations. He sluiced out a tail-race and wheel pit this winter (1869) in a hard gravel soil, in a very short time; he placed a sluice (by nailing three boards together, forming a bottom and sides, and open at top) at the lowest end or discharge of the race, and ran the water down the intended route, and through this first length of sluice into the river. He then commenced at the end of this length and loosened the earth with a pick or crowbar to the required depth and width, and the water, rushing down from the unbroken surface above, swept the earth and stones down the spout into the main stream. When thus excavated to the proper distance, another length of sluice was added, a little straw stuffed under the end, and each side blocked up with stones to direct the water into the end, and another length loosened and sent through the sluice, as before. This process was continued until the race was completed, and the foundation for the wheel-house reached; and here the benefit of Mr. Whipple's Californian experience was more conspicuous than in digging the race, because the depth of excavation here was so great that the earth would have had to be wheeled to get it out of the way, and this is always a tedious process; but he sent all the earth and moderate sized stones down the sluice and river in an incredibly short time; by shifting the chute of water upon different parts, and loosening and throwing out the larger stones, thus giving the rest a proper facility of entering the sluice, it was swept down stream by the water, requiring only a little assistance with a hoe or shovel when large or flat stones would incline to stop and obstruct the passage.

Any person who does not comprehend the assistance which such a sluice renders, to enable the water to carry away stones and gravel, may satisfy himself by trying to shovel or push such gravel and stones on top of their natural bed, and then try to shovel the same, or push them along in such a sluice; or let him try to shovel potatoes from the top of a bin, and then try shovelling them on a wooden floor.

Care should be taken to place each length of sluice at the full and proper depth at first, and also to give each length an equal and sufficient fall to insure the requisite rapidity of current. By Mr. Whipple's experience, two inches of fall in one length of board, or twelve feet, will carry along stones the size of potatoes, or a man's fist, while three or four inches fall to each length will roll along stones as large as a man's head. Of course, a sufficient supply of water is necessary in all such operations.

Pressure of Water.

When water is confined in a tank or flume, it presses equally on every part of the interior of the bottom or sides, with a force proportioned to the height of the water.

Thus, a flume or tank, four feet square, and filled with water one foot deep, will sustain a pressure of 62½ lbs., or 1000 ounces, the weight of a cubic foot of water, on every superficial foot of the bottom, or 4×4=16 feet × 62½=1000 lbs. altogether. But the pressure against all the four sides, which makes an equal area of 16 feet, will be only half that amount, viz., 31¼ lbs. per foot, or 500 lbs. altogether; because that although the pressure is the same at the lower edge near the bottom as upon the bottom itself, yet that pressure diminishes upward to nothing at the surface of the water, and for that reason the pressure is intermediate, and just half that upon an equal area of the bottom. Now, suppose we let the flume fill three feet deeper, making four feet of water, then the extra three feet are added to the other, and the pressure on the bottom will now be 4000 lbs. instead of 1000, and the whole additional three feet deep and 3000 lbs. pressure must be added to the former 500 lbs. of lateral pressure, outward, against the lower foot deep of the side planking, thus making 3500 lbs. against the lower 16 superficial feet of side planking, while, as before, the pressure against the foot deep at the surface will only be the same 500 lbs. To find the amount of pressure against one whole side, the computation, as with one foot deep, must be taken at the middle. Thus the area of one side being now the same as that of the bottom, 16 superficial feet, the whole pressure of the water against any of the sides will be just one-half of that upon the bottom, or half the depth, 2×1000= 2000 lbs. The pressure against the inside of a flume or penstock of any depth, and the whole weight of water contained in it, can be computed in the same way.

If we fill this same flume up to 16 feet deep, instead of 4 feet, the pressure and weight upon the bottom will be four times as great, and the lateral pressure and strain will be increased in the same proportion. But we might deck the flume over with plank, at the surface of the four feet of water, and instead of filling the whole flume to 16 feet deep, insert a tube into the water through the deck, and reaching up to 16 feet, or any other height, and by filling this tube with water, the pressure on all parts of the interior of the flume, under the deck, will be the same as if the whole flume were filled to the height of the tube; and the velocity of the water issuing from any part of the flume will also be the same, if the tube be sufficiently large to furnish the quantity discharged, and be kept full.

In this case the water in the flume will press upward on every part of the under surface of the deck with the same force that the column of water contained in the tube will press downward upon the water immediately under its lower end; and if a hole be bored through the deck, and a tube of proper shape to direct the water be inserted in the hole, the water will be shot up as high as the surface of the water in the tube, except a slight reduction caused by the friction through the tube and the atmosphere.

Circumstances frequently occur in the construction of mills, especially where the head of water is high, which render it necessary to construct the flume on this principle. Less strength is required in the framework, less weight thrown on the supporting timbers, and it may be placed under a floor comparatively low, and supplied by a tube at any angle, from the outside of the building. The above remark, that the frame of such a flume requires less strength than if carried up the full size and whole height, needs some explanation, as the pressure of the water upon every square inch or foot of the interior, is the same in both cases. But the area of surface within the decked flume and tube is less, and diminishes the whole strain on the framework in like proportion, and the timbers are stronger in proportion as they are shorter. There is also an excessive strain on the interior of a close flume when a gate is shut, which must be relieved by a ventilating tube carried up from the deck to a sufficient height.

Velocity of falling Water.

When water is allowed to fall freely from any height to the ground, it obeys the same laws, and acquires the same velocity as a solid body of the same weight would acquire in falling through an equal distance. This velocity increases in a regular ratio, as the distance of the fall is increased, and is the *theoretical* velocity ascribed to water issuing from a flume, or penstock, under the same head as that fall. The *actual* velocity of water issuing through a hole cut in such a flume falls far short of that velocity, as has often been proved by experiment; the actual velocity being to the theoretical as 10 is to 16. But the full theoretical velocity ascribed to the water issuing under any given head can be obtained, and even exceeded, by a proper construction of the spout or chute through which the water issues; and as this subject is of great importance to the millwright, we shall try to explain it here.

When such an orifice is opened in a flume, the water rushes toward the outlet from every part of the interior; here the different currents all meet, and their opposing forces mutually obstruct and impede each other in their progress toward the outlet, and thus diminish the velocity of the column of water as it enters the orifice. But in passing out through the orifice, these opposing currents are equally balanced; and being thus united and the sum of all their forces tending in the same direction, the velocity of the issuing column is increased, and in consequence of this increase of velocity, its diameter is diminished; and it is found by measuring the discharging column, that its diameter at the smallest part is to the diameter of the orifice as 10 is to 15 or 16, accordingly as the edge of the orifice is thicker or thinner. By attaching a short tube, of the same bore, to the orifice, the effect of the opposing currents inside is partially counteracted, and the discharge through the tube will be as 13 to 16 of the theoretical velocity; but by widening out the end of the tube next the interior, to the shape of the column issuing from the hole, the full theoretical velocity of the water is obtained, which, as already intimated, is equal to the velocity attained by a heavy solid body falling freely through the distance from the surface of the water to the orifice. The conical part of the tube or spout should only extend half its diameter from the flume, which is the point of greatest contraction in the liquid column, and the area of the wide end should be to that of the narrow end as 16 to 10, like the issuing column.

When we consider that the power obtained from water increases, not as the simple increase of velocity, but as the *square* of the velocity, and that, to *double* the velocity through any given spout yields *eight times* the former amount of power, the importance of this subject will be seen at once. To explain this, it is necessary to consider that the double velocity gives $2 \times 2 = 4$ times the power from the same quantity of water, but a double velocity passes through a double quantity in the same time, therefore it must be doubled again, which makes $4 \times 2 = 8$ times the power.

To make this plainer, suppose a water wheel is only capable of driving a part of the required machinery, and more power must be added; if we can by raising the head, or otherwise, impart a double velocity to the water passing through the same chute, eight times the former power will be imparted to the wheel; and this, if all the parts be strong enough, will drive more machinery than if seven other wheels were added, and each supplied with an equal quantity of water, and at the same velocity as the first. Here, we see, that by doubling the velocity of the water, only twice the former quantity is used to impart eight times the power, while it would require eight times the quantity of water at the original velocity to obtain the same result; and when we take into account the extra expense, room, and friction which the latter arrangement of eight separate wheels would involve, we can appreciate the importance of increasing the velocity of the water applied to any water wheel or turbine; and the importance of forming all chutes and delivering issues, so as to assist instead of retarding the required velocity. See remarks on this subject in the chapter on *Central Discharge Wheels.*

Tables.

We subjoin a few Tables which will be useful in estimating the powers of water, under different circumstances, and in making the necessary calculations to adapt the power and machinery to the intended purpose. As a millwright cannot be expected either to commit these tables to memory, or to have them always at hand, the following short rules, which are easily remembered, will be found convenient, and sufficiently accurate for practical purposes:—

1. To find the velocity of water under any given head.

RULE.—Multiply the square root of the head in feet, from the middle of the gate to the surface of the water, by 8, and the product is the velocity in feet, per second.

Examples.

Ex. 1.—Required the theoretical velocity of water, under 16 feet head?

Ans.—Square root of $16=4^2\times 8=32$, the velocity in feet per second.

Ex. 2.—Required the theoretical velocity of water, under 25 feet head?

Ans.—Square root of $25=5^2\times 8=40$, the velocity in feet per second.

Ex. 3.—Required the theoretical velocity of water, under 49 feet head?

Ans.—Square root of $49=7^2\times 8=56$, the velocity of feet per second.

2. To find the quantity of water discharged.

RULE.—Multiply the area of the gate in feet, by the square root of the depth in feet, and that product by $5\frac{1}{10}$, the product is the quantity in cubic feet, per second.

Examples.

Ex. 1.—Required the quantity of water that will pass per second through a gate 2 feet square, under 9 feet head?

Ans.—Area of gate $2\times 2=4$. Square root of $9=3^2$ $4\times 3=12\times 5\frac{1}{10}=61\frac{2}{10}$ cubic feet.

Ex. 2.—Required the quantity discharged per second through a gate of 2 feet superficial area, under 25 feet head?

Ans.—Area of gate 2. Square root of $25=5^2\times 2=10\times 5\frac{1}{10}=51$ cubic feet.

Ex. 3.—Required the quantity of water that will pass per second through a gate 1 foot square, under 36 feet head?

Ans.—Area of gate 1. Square root of 36 $= 6^2 \times 5\frac{1}{10} = 30\frac{6}{10}$ cubic feet.

The *first* rule gives the *theoretical* velocity sufficiently accurate, as will be seen by comparing the foregoing results with the table.

The *second* rule gives the *actual* quantity discharged, as shown by the same table.

Table of Velocity of Water and Quantity Discharged under Different Heads.

	THEORETICAL.		REAL.			THEORETICAL.		REAL.	
Head in feet.	Velocity per second in feet.	Cubic feet discharged per minute. Area of orifice 1 inch.	Velocity per second in feet.	Cubic feet discharged per minute. Area of orifice 1 inch.	Head in feet.	Velocity per second in feet.	Cubic feet discharged per minute. Area of orifice 1 inch.	Velocity per second in feet.	Cubic feet discharged per minute. Area of orifice 1 inch.
Ft.	Feet.	Cu. feet.	Feet.	Cu. feet.	Ft.	Feet.	Cu. feet.	Feet.	Cu. feet.
1	8.02	3.34	5.09	2.12	23	38.46	16.02	24.42	10.17
2	11.34	4.73	7.20	3.00	24	39.29	16.37	24.95	10.39
3	13.89	5.79	8.82	3.68	25	40.10	16.71	25.46	10.61
4	16.04	6.68	10.19	4.24	26	40.89	17.04	25.97	10.82
5	17.93	7.47	11.39	4.74	27	41.67	17.36	26.46	11.02
6	19.64	8.18	12.47	5.19	28	42.43	17.68	26.94	11.23
7	21.22	8.84	13.47	5.61	29	43.18	17.99	27.42	11.42
8	22.68	9.45	14.40	6.00	30	43.93	18.30	27.90	11.62
9	24.06	10.02	15.28	6.36	31	44.65	18.60	28.35	11.81
10	25.36	10.57	16.10	6.71	32	45.37	18.90	28.81	12.00
11	26.60	11.08	16.89	7.04	33	46.07	19.20	29.25	12.19
12	27.78	11.57	17.64	7.35	34	46.77	19.49	29.70	12.38
13	28.91	12.05	18.36	7.65	35	47.46	19.77	30.14	12.55
14	30.01	12.50	19.06	7.94	36	48.12	20.05	30.56	12.73
15	31.06	12.94	19.72	8.22	37	48.78	20.33	30.98	12.91
16	32.08	13.37	20.37	8.50	38	49.44	20.60	31.39	13.08
17	33.07	13.78	21.00	8.75	39	50.08	20.87	31.80	13.25
18	34.03	14.18	21.61	9.00	40	50.72	21.13	32.21	13.42
19	34.96	14.57	22.20	9.25	49	56.07	23.79	35.04	14.87
20	35.87	14.95	22.78	9.49	65	64.08	27.61	40.05	17.26
21	36.75	15.31	23.34	9.72	100	81.00	34.00	50.62	21.25
22	37.61	15.67	23.88	9.95	144	97.02	40.80	60.61	25.50

Measuring a Stream of Water.

The following table gives the number of cubic feet of water passing over a wier per minute, for every inch in length of wier, from 1 inch up to 18 inches deep.

RULE.—Multiply the number corresponding with the depth, by the number of inches in length of the wier, which gives the number of feet passing per minute. Then multiply the number of feet by 62½, and that product by the head in feet, and divide by 33,000, which will give the horse power.

Table of Quantity of Water passing over a Wier from 1 to 18 inches deep.

No. of inches deep.	No. of cubic feet passing per inch in length.	No. of inches deep.	No of cubic fe passing per in in length.
1	0,403	10	12,784
2	1,140	11	14,404
3	2,095	12	16,758
4	3,225	13	18,895
5	4,507	14	21,170
6	5,917	15	23,419
7	7,466	16	25,800
8	9,122	17	28,250
9	10,882	18	30,706

Example.

1. Required the power of the water passing over a wier 10 inches deep and 36 inches long, with 24 feet head or fall?

```
         12,784 cubic feet passes over 1 inch wide.
             36 length of wier.
         ------
          76704
         38352
         -------
        460,224
            62½ lbs. weight of water in cubic foot.
        -------
         920448
       2761344
         230112
       --------
      28764,000
             24 head in feet.
      ---------
      115056000
      57528000
      ---------
33000)690336,000(20,919 3/11 horse power.—Ans
      66000
      -----
       303360
       297000
       ------
         63600
         33000
         -----
         306000
         297000
         ------
           9000
```

CHAPTER V.

WATER WHEELS—THE UNDERSHOT.

THE undershot wheel, Fig. 22, the first and oldest method of obtaining a motive power from water, is now seldom used in the United States, being mostly super-

Fig. 22.

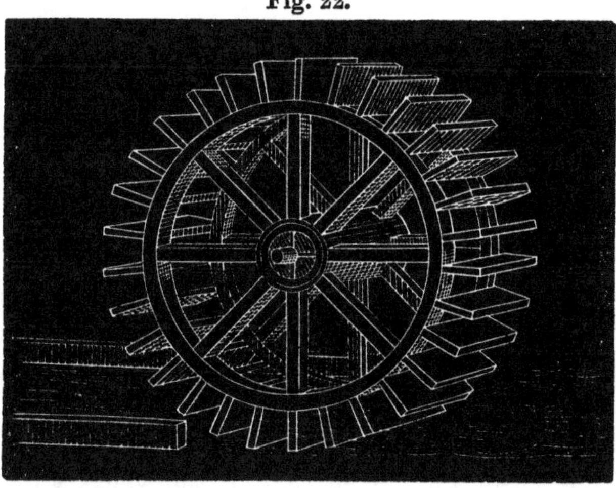

seded by others, more convenient for gearing machinery, and occupying less room. It therefore requires but a brief notice here.

It is simply a horizontal shaft, with arms attached, projecting to a length proportioned to the intended diameter of the wheel; float boards are fastened across the ends of these arms, parallel with the shaft, which is placed across the issuing stream, into which these float boards dip, and are carried round by its force. If a greater number of float boards are required than there are arms, rims composed of felloes, are placed around the outer ends of the arms into which starts are framed, and the floats are attached to these starts. Care must be taken to form the water-way so that the floats will just pass through clear, without allowing the water to waste, and also to allow the water to pass away freely when it has spent its force upon the float; and also to place the floats sufficiently close to each other to prevent the water passing between without spending its force, and not close enough to cause the water striking against the back of one float to recoil back upon the front of the one behind it.

The paddle wheel of a steamboat is a sample of an undershot of very large diameter; the flutter wheel, so common in old saw-mills, will represent a small one.

The power of this wheel depends upon the quantity and velocity of the water acting upon its floats, and as so many other wheels come under the same rules, we would refer the reader to the table in the preceding chapter, for the velocity due to any particular head or fall, and also to the same tables for the quantity passing through a given aperture; merely explaining here that the millwright must be particular to proportion the diameter of the wheel and the gearing to the velocity of the water due to the particular head, as found in the table. Here good judgment is required to obtain the best result, as under some circumstances the wheel may move with two-thirds of the velocity of the water, while in others it should only move at one-third of that velocity. Suppose it is a large undershot, where considerable gearing is required to connect it with the work, as in a grist mill, then the wheel should be geared to move at about one-third the velocity of the water; but when a wheel of small diameter is used, like the saw mill flutter wheel, with the crank on the same shaft, dispensing with gearing, then the wheel may move with two-thirds the velocity of the water, and it will work more steadily and give a better result.

To understand this better, suppose the wheel to be geared to a grain elevator; hoist the gate, and let on the water, with the elevating buckets all empty. The buckets will move up with great speed, and the floats of the wheel will travel nearly as fast as the water, but no grain will be carried up. Now stop the machine, and fill grain into all the ascending buckets until it exactly balances the power of the water, and it will stand still with the water all on, but no grain is elevated. These are the two extremes, and the best result lies somewhere between. Now take out half of the grain from each bucket, and try it again; the cups or buckets, and the wheel will move at about half of their velocity when empty, and continue to carry that half load at that half speed; put less grain in each cup and they will travel faster with the lighter load; put in more, and they will travel slower, with a heavier load. Try the machine with different loads and velocities, until you find the rate at which it will carry the greatest quantity in a given time, and then you have the right speed to work it at. These conditions apply equally to any other kind of wheel, and all other kinds of machinery.

No invariable rule can be given. Old millwrights, who have experimented much on this kind of wheel, as they were much used in their time, differ widely in their conclusions, and present experience shows that the maximum velocity for working them still maintains as wide a range as their various theories gave it, that is, from one-third to two-thirds of the velocity of the water.

As great a difference obtains with regard to the *power* of these wheels, some experiments making them capable of raising one-third of the quantity of water expended on them, to the same height, while others make the quantity only $\frac{4}{27}$ths, or even one-eighth. This is a natural consequence of the great variety of circumstances and situations in which they are placed, and the diversity of opinion among millwrights themselves.

There are many situations where this wheel may still be used to advantage, and be preferable to any other. Where there is a large quantity of water, and little head or fall, a given amount of power can be obtained from an undershot easier and cheaper perhaps than by any other wheel. By using large float-boards, a great quantity of water may be applied to give the power required, without the expense of erecting high dams or expensive flumes and water-ways required for other kinds of wheels, the quantity used making up for the low head and little velocity obtained.

We remember seeing a wheel of this kind about forty years ago, at Lachine, on the island of Montreal. The priests and nuns of the Seminary owned all the island, and all water-power connected with it (by a grant from the French Crown, before the treaty ceding Canada to Great Britain was signed), and the mills were mostly driven by wind. This element is proverbially unstable, and the mills driven by it partook of the same characteristic. But a cunning cannie Scotchman built a mill at the Lachine Rapids on the St. Lawrence River, that was to be more reliable. The priests being inexorable in their monopoly of the water-power, Mr. G. built his mill ostensibly for horse-power, but did not run it long in that style. It stood as close as compatible with safety, to the Rapid, and Mr. G. placed a large undershot, or flood-wheel out in the current, taking care to have it beyond the low-water mark; this wheel he connected with the mill machinery by a long shaft and coupling irons. To start the mill, the clutches of the couplings were thrown in gear, and to stop it they were thrown out again; the water-wheel revolved continually, night and day, summer and winter, without intermission. Its motion was so constant and uninterrupted that the French Canadians made it a by-word or adage; any person who talked continually, to the annoyance of others, they say their tongue was *comme le Moulin Lachine* (like Lachine Mill), that never stops. The Seignors prosecuted Mr. G., and a long lawsuit was the result, which ended in Mr. G.'s favor.

There are many situations where such rapids occur, and where the stream is large, and other circumstances favorable, quite an amount of machinery might be driven. We never erected such a machine in a rapid, but have frequently driven light machinery on the rapids of comparatively small rivers, by a kind of screw-propeller, resembling a large grain-conveyer, laid lengthwise in the stream. This revolves more rapidly than the undershot, and the required speed can be obtained with less gearing. The speedier motion also admits of using this kind where the current is less rapid than the undershot will admit of, by lengthening the shaft—to the length of a mast, if you like—and adding more oblique screw-paddles, the power can be multiplied to a great extent. We have seen a grist mill of three run of stones driven by such a wheel, in the Genesee River, somewhere between Rochester and the Alleghanies. This was before the Genesee Valley Canal was made, and the river boats passed up and down through the rapid in which this screw-wheel was submerged. But this is a digression, and these wheels will be described under the name of Spiral or Screw Flood Wheels.

The great objection to using the undershot in a cold climate, is the trouble and expense of keeping it clear of ice. This applies equally to the overshot, and requires that it should be inclosed in a tight wheel-house which is heated by fire, steam, or smoke. A common and convenient way is to have a small steam boiler fitted into the top of the mill stove, from which a pipe runs to the bottom of the wheel, and discharges the steam there; the steam will ascend about the wheel when it is idle, and circulate with it when in motion, and if the wheel-house is close, will keep it free from ice. This plan has the objection of keeping the roof and whole interior of the wheel-house wet with condensed steam, which causes it to rot.

In the northern portions of the United States and the adjoining provinces, a stove or fireplace is placed in or near the wheel-house, discharging the smoke and fire into the wheel, which circulates it by its motion, like the steam in the other case. This is preferable to the steam, but requires that the wheel-house should be separated from the mill by a tight wall, that the workmen may not be annoyed by the smoke. It is not essential that the wheel-house be perfectly tight in this case, because the smoke and hot air carried in with it do not condense like steam, and while the supply is kept up smoke will be seen issuing from every crevice, effectually preventing the entrance of cold air. But steam condenses so rapidly that there is nothing to ooze out through the cracks, which, as well as the tail race, must be tightly closed. The steam supplied by the mill stove has the advantage that it is sure to be well fired up in cold weather, while the wheel-house stove or furnace is liable to be neglected in a busy time.

The undershot wheel may be geared to machinery, either by placing cast-iron cogged segments around the wheel itself, or by extending the main shaft to a suitable length, and placing another wheel on the end upon which to place the gearing, and driving another shaft from this by a pinion. See directions for gearing the overshot.

CHAPTER VI.

THE OVERSHOT WHEEL.

The overshot is the most powerful of all water wheels, when tested by the power yielded from the same quantity of water; and this will apply to every height of fall, whether high or low, where the overshot can be used. An idea is very prevalent among millwrights and mill owners, that the overshot is not suitable for a moderate fall, which idea only requires testing to show it to be erroneous. Mills may be seen in many parts of the country, driven by various kinds of wheels, and working on a short allowance of water, which might have enough and to spare, if these wheels were used. Many of the most intelligent millwrights are aware of this fact, but almost every millwright of note is either a patentee, or in some way interested in the manufacture or sale of some other kind of wheel, which he recommends. A few of these are excellent wheels, and give results nearly approaching the overshot; while a greater number bear a closer affinity to the undershot.

The superior power given by the overshot is due in part, to the more economical application of the water,

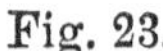
Fig. 23.

every particle of which is applied to the best advantage for moving the wheel, while with others, there is more or less waste by leakage and friction; but it is principally due to the fact that the water acts by its gravity during its whole descent upon the wheel, while it is only by its acquired velocity that it is made to act on other wheels. This velocity is found in practice to fall far short of that ascribed to it by theory. This, with the great difficulty of applying and arresting the whole acquired velocity of the water, accounts for the difference. A well-constructed overshot will raise nearly as much water as is applied to it, to the same height as the fall, provided the fall be reckoned from the point where the water enters the buckets to the point where it spills out of them, while some other wheels will only raise one-sixth or even one-eighth of that quantity. A difference of opinion exists among millwrights as to the proportion the height of the wheel should bear to that of the fall, some contending that the wheel ought to be higher than the fall, and others, that it ought to be lower. A favorite idea with the former is, that the wheel should be so much higher, that the water may be applied one-third down from the top, or at 10 o'clock on a dial plate. This, however plausible it may appear in theory, is not found to hold good in practice, because the increased size of the wheel which this requires makes it of a larger circle, which causes it to spill the water higher, and wastes more of the effect of the fall than the smaller wheel. Besides, the water on entering the buckets requires some time to settle before it bears its full weight upon the wheel, and when shot in at the top it has time to subside while passing nearly horizontal, and is ready to exert its whole force when the descent commences; whereas, when let on at the point corresponding with ten on the dial, the descent being already abrupt, a slight loss of power occurs while the water is subsiding. Add to these considerations the extra cost, weight, and room required for the large wheel, and the fact that it requires more gearing, as its revolutions are slower, and the odds will be in favor of the smaller wheel.

There appear to be only two considerations that would justify making the wheel lower than the fall; one is, where there is more fall and water than is required for the work intended—and the other, where the fall is so high that the wheel would be too large to be convenient and durable. It is therefore recommended that where the fall is permanent, the wheel be made only so much lower than the fall as to admit of shooting the water properly into the buckets, which will be considered in treating of the velocity at which an overshot ought to move. But where the fall is variable, with a pond to draw upon, then the wheel should be low enough to get the benefit of the pond for gathering water. If care be taken to shoot the water properly into the buckets, the head above the wheel is not lost, but acts upon the wheel the same as upon the best constructed undershot, and every two feet of that head will produce an effect about equal to one foot in height of the wheel.

There is considerable difference of opinion among millwrights as to the rate at which an overshot ought to move. A favorite speed with many old millwrights is three feet per second. This would appear to be too slow under any circumstances; it causes too much gearing to intervene between the wheel and the work, and too much water to be loaded upon the wheel, thus producing a needless extra strain and friction upon every cog and journal. If the same wheel were geared to move at six feet per second, instead of three, only one-half of the gearing would be required to give the machinery the same speed, and the wheel would require but half the capacity for water, while all the various parts would be lightened in the same proportion.

This, it is evident, would produce the same effect with the same water, and the wheel would work much freer and with less strain and friction, there being only half the quantity of water in each bucket, with double the purchase on the machine driven. Thus, the same water wheel that drives a certain amount of machinery when moving at three feet per second, will drive double that amount when geared at six feet per second, but requiring double the quantity of water to be discharged upon it, as each bucket remains only half the time under the stream; hence, the head over the wheel must be more in the latter case, to give the water a velocity greater than the wheel, which in every case it ought to have, in order to shoot it properly into the buckets.

Many good millwrights err in this particular, and some even take the trouble to interpose an obstruction between the flume and the wheel, to destroy the velocity of the water before it reaches the wheel, under the mistaken idea that it thus produces more effect. In this case, the front of the bucket strikes with its flat surface against the stream, and carries it in a broken mass before it for a considerable distance before it enters the buckets at all, thus not only losing a portion of the fall, but interposing an actual resistance to the motion of the wheel so long as the water lies against the front of the bucket.

The principal argument in favor of running a wheel so slow is, that by remaining so much longer under the stream the bucket has more time to fill; as an old millwright here expresses it, he "wants thirty yards of the stream on the wheel at once." This is a fallacy, as it cannot be done without interposing too much complicated and expensive clock-work between the wheel and the work.

This misconception originates in the idea that the slower any body descends by the action of gravity the more force it exerts on any opposing obstruction. This also is a fallacy, as can be proven when the time occupied by the descent is taken into account, which it must always be to arrive at true conclusions. The fact is, that the action of gravity is not interrupted by motion at all, no matter what the velocity may be; a jet of water or a cannon ball projected from a precipice horizontally, over a level plane, will strike the ground at the same instant that water, or another ball, dropped perpendicularly from the same height, at the same time, will strike the ground. When this is understood and admitted it will be easy to see that there is no real loss of power in running an overshot at considerable speed, excepting only the head of water necessary to obtain a velocity greater than that of the wheel.

We built a large overshot, many years since, which was geared to run nine feet per second, with a variable head from four feet downwards, which we have never been able to beat since; and would remark that the larger the wheel the faster it may be run; and a very small diameter ought not to be run faster than four feet and a half per second. In every case the water should get away freely from under these wheels. The millwright, in making calculations for the construction of these wheels, should consult the table of the velocity of water, and allow sufficient head above the wheel to give the water sufficient velocity over that of the wheel to insure its free entrance into the buckets.

Having considered the theory and principle of action of the overshot wheel, we will now proceed to detail its mode of construction.

Having taken all the levels from head-water to the tail-race (for directions, see article on levelling), making due allowance for head over the wheel, and free discharge below, and decided on the diameter and size of the wheel, you will procure a sweep-rod a little over half the diameter of the wheel for a radius; stick a scratch-awl through one end of this rod, and into a clear floor of sufficient size; then measure off half the diameter of the wheel on the rod, from the first scratch-awl, and insert another awl. Move this last round until you describe the outside circumference of the wheel upon the floor; then move the outside awl back on the rod to the width or depth of the rim, and describe another circle there. Step the outer circle and divide it into as many parts as there are arms and felloes in the rims.

Then take a true straight-edge of sufficient length, or if that is not handy, a fine chalk-line will do, and draw a scratch from each of these divisions to the centre. This last mark will describe the centre of the arms. Place a thin board upon the rim draft, equally between two arms, and describe the two rim circles upon it; now extend the scratch describing the centre of the arms across this board, and the rim pattern will be marked out. Proceed to lay out the gains or notches for the buckets, on the pattern, before you remove it, because these must be laid out very exactly, in order that the part of the buckets divided by the joint at one end of the pattern may correspond exactly with the part at the other end. The *start*, or first part of the bucket, is laid by the radius from the centre. In laying out the *shoulder*, or outer part, the millwright must use his own judgment, as it must be varied according to the depth of the rims, distance apart of the buckets, and the quantity of water to be used, taking care to leave room for the free inlet and outlet of water; and at the same time bearing in mind, the greater the angle of the buckets the longer will they retain the water at the bottom, and so work to better advantage.

The pattern now requires the arms to be marked upon it; to do this extend a mark across each end of the pattern parallel to the centre mark of the arm, and half the thickness of the arm from that mark, as one-half of the rim pieces must be cut off by that mark, if the rims are to be double, to allow them to go in between the arms. The other half must be left of full length to meet half way between the arms. On these rims the arms require to be marked at the centre, and to lay these out you require the same mark upon the pattern. To get this, find the centre of the pattern, and draw a scratch from this to the centre of the wheel, the same as for the other arm marks, and mark the whole size of the arm by this centre mark, one-half being on each side of, and parallel to it. Now take up the pattern and work it off, and this will answer for laying out all the rims for both sides of the wheel. For laying out one side of the bucket gains, the pattern must be marked on the other side—to transfer the marks, square across both edges of it at the marks, and bore a small brad-awl or gimlet through both the outer and inner corner of the angle—place the square from these holes to the scratches on the edge, and draw the marks required.

There are several ways of putting these rims together; some make the rims of single felloes: these are notched at the centre to go into a slit in the end of the arms, which are then bolted through; the ends of the felloes are drawn together by joint bolts; these are round bolts with a nut on each end, and are bored in like dowels with a mortise through the side into the hole, for screwing on the nuts. When the ends are drawn tightly together they are fastened by an iron clamp outside and inside and bolted through. This is a common way of making the rims, but it is preferable to make them of two tiers of plank, the inner one say three inches thick, the outer two inches. Sink the gains for the buckets, and one at the centre for the arm, one inch deep, which will leave two inches of whole wood in the three inch pieces, the notch for the arm being on the same side as the others; make a slit in the ends of the arms the depth of the rim, two inches wide, and two inches from the outer edge of the arm. Slip the notched part of the thick pieces down into these slits and bolt them, which will complete the inside tier. Cut the ends off the two inch rims to the arm marks on the pattern, and put them in between the arms, and they will just fill it out even, making it smooth outside. A portion of the arm will project inside, to which the opposite buckets must be fitted. These will break joints, and when well pinned and bolted together will be strong and durable. If a greater thickness of rim is desirable, to put on segments or the like, another tier of any thickness may be put on the outside, breaking joints with the others, and of course adding to the strength.

The rims thus made of two thicknesses will be of half strength at the weakest part, and if of three tiers there will be two-thirds whole wood at the joinings. It is of importance in making large wheels to have them strong, without useless weight in any part, and free from weak points. When the rims are made of a single felloe, fastened with iron as described, the joints are all weak points, and when the wheel is in motion there is a great and continually changing strain upon these points; when the joint is below, the whole weight of the wheel and water is pressing it together; but when it is at the top the same strain is drawing it apart; and as this reciprocating strain is constantly acting upon the joints while the wheel is in motion, it is apt to work some of the iron fastenings loose, in the course of time; and when a wheel begins to *work* at the joints it soon goes to destruction. Single rims are sufficient for a wheel of small diameter, as the arms being short are stiff enough to keep the whole in place; but in a large wheel, the arms above and under the shaft, which act endways, have the whole weight of wheel and water to sustain, as the arm when horizontal only keeps the wheel in shape, being so long and slender that its own weight would spring it down at the ends if not held up by the rims. It is obvious, therefore, that the large wooden arms act partly on the suspension principle, as well as those composed of iron rods, and that every large wheel depends on the strength of its own rims for stability.

For all very large wooden wheels I would recommend the following plan, the result of a rather extensive experience, as the most economical and durable: Make the arms small, but as many in number as can be properly attached to the shaft, in order to have the rims supported at as many points as practicable; then make the rims like the rim of a large spinning wheel. To do this good wood is necessary, such as oak, elm, or ash; it should be sawed about two inches thick, and of a width equal to the intended thickness of the rims; a circle or mould should be made equal to the inside circle of the rims, and long enough for one length of the stuff. Steam enough of the pieces to make the first tier all around the rims, and fasten them upon the mould until they are dry enough to retain their shape; make a shoulder upon each arm, at the inside circumference of the rim, by cutting away a portion of the inside of the arm and leaving a piece outside to bolt through. Now take the pieces off the mould, and lay out a gain (by the pattern or draft) in the edge of the pieces to fit the projecting portion of the arm, and half an inch deep. The parts may now be put together by placing them around the shoulders on the arms; bring the projecting *lug* of each arm into its corresponding notch, and drive a spike through into the shoulders to secure them in place. The first tier being now completed, there is no further need of the mould. Take the pieces from the steam-box and bend them on to their places by griping down the ends and driving a few nails here and there to keep them even, taking care to place the nails where they may not interfere with the bolt holes. Thus proceed to put one tier above another, taking care to break joints, until the rims are of the required depth; then bolt them together with plenty of small bolts, and fasten a strong piece of iron upon the inside of each arm, and bolt the rims through this and the projecting portion of the arm, and the rims will be complete.

To put in the buckets, lay out and cut gains or recesses around the inside of the rims the same as when made of felloes. I have used the following plan, which I can recommend as being easier and cheaper, and equally compact and durable: Take thin inch boards, and make a pattern exactly corresponding with the solid piece left between the bucket gains; lay this pattern upon the boards and mark out as many pieces as are required for both sides of the wheel; saw them out and dress them off to the pattern, taking care to have the grain of the wood run parallel with the long point of the bracket; gauge and dress off the ends of all the buckets to a size, and they are ready to put in. Commence by placing a bracket on each side, and nail it with about five small nails in each; place the two pieces composing a bucket, and nail them securely together at the corner; place and secure another pair of brackets, and another pair of bucket pieces, until the whole is completed. When you come to the iron straps at the arms cut away a portion of the bracket to fit the iron, and a piece out of the end of the bucket to fit the same. Next begin by placing a piece of the lining and nailing it securely to the rims, and also to the inner edge of the bucket, which you must be careful to do with each piece as you proceed, as it is difficult to strike the edge of a bucket after a number of them are covered.

We have made a number of large overshot wheels in this way, and they are both *light* and *strong*—two qualities of great importance. This plan also involves a much smaller waste of material and labor, which is another important consideration.

I have unwittingly detailed the methods of constructing the outer circumference of the wheel before describing the shaft and centre portions by which it is supported; which seems to indicate that I am more familiar with *making* a wheel than *describing* one on paper. But as there is no danger of any one committing the same blunder in constructing a wheel, the error is comparatively harmless.

We will proceed to describe the process of making the shaft. An iron shaft requires but little description, as it is generally ready made to order. The first consideration is to get suitable flanges upon it to fasten the arms upon. These may have sockets like those in use for steamboat paddle-wheels; but these are not absolutely necessary, unless the power of the wheel is to be conveyed through the extended shaft, which is sometimes requisite. The most economical mode of transmitting the power is by placing cogged segments around the circumference of the wheel, to work in a pinion upon another shaft, to which it gives motion. As this motion is taken from the swiftest moving part of the wheel, it is comparatively rapid, and but little more gearing is required to connect it with any machine to be driven. This pinion shaft may run in almost any direction, horizontal or perpendicular, by using spur, or bevel segments. When motion is taken from the wheel in this way the shaft does not require to be so strong, as it has only to support the weight and keep the wheel in position, there being no twist or other strain upon it. In this case all the central flange required is a plain round disk of sufficient size to bolt the arms securely upon, and of sufficient strength at the centre to be keyed firmly on to the shaft.

To lay out these arms, make a draft of the size of the shaft and flange upon the centre of the wheel draft; then mark out the number, size, and position of the arms, and by one of these arm-marks make a pattern for laying them out, taking care to butt the ends solid against the shaft. After the arms are bolted on, fit a wedge-shaped piece in between, of sufficient size to fill the space for some distance past the disk. Put these wedges in with hot pitch, or white lead paint, and bore a hole through, half in the arm and half in the wedge, and insert a pin. This will bind the whole centre into a solid hub, which is strong and secure. Such a shaft is preferable to a wooden shaft and iron journals; and these, with the necessary bands and workmanship, will frequently cost more than the cast iron shaft. This consideration, coupled with the fact that the wooden shaft is liable to rot and other casualties, while the iron one is sure to last out the rest of the wheel with little deterioration in value, makes the iron preferable. Old broken or worn-out steam engine shafts are sometimes to be had for little more than the price of old iron, and answer well.

When a wooden shaft is to be made, oak or tamarac are the best woods, while black ash and pine are very good. And here I might observe, that tamarac or *larch* is the very best wood for the wheel itself, being strong and durable, and, like pine, not liable to waste away outside, or *rust*, as it is technically called, as all hard woods are. To make the shaft, be careful to get a sound stick, and large enough; if you can get it to a saw-mill, it will save some hard work. Cut off the round log an inch or two longer than the length for the shaft; sink in a piece of iron with a countersunk centre, in the middle of each end, and hang it up a convenient height upon two iron or steel points in posts or other fixtures, in such a position that you can pass a belt around a shaft, and around the log, to give it motion, and turn it off true, the whole size and length. Mark the shoulders of the gudgeons and turn them off to the size and taper of the wings; mark also the mortises for the arms, with another scratch outside of the mortise for the dovetail, about half an inch, and one inside the mortise, for keys, about three inches; turn the shaft round and scribe all these different marks. Now measure the size of the shaft and describe it on the floor, in the centre of the wheel-draft, and divide it and lay out the number and size of the arms, to correspond with the wheel. By this section of the shaft make a pattern for the thickness of the tenon on the arm; attach a piece of circle agreeing with the outside of the shaft across the top of this pattern to gauge the depth of mortise, and to insure making them plumb and true, and another similar gauge for the dovetails. Make the mortises before you take down the shaft. You may now make a pattern of the back end or wings of the gudgeons, as they are too heavy to handle, and lay out and work the recesses by it. It will save some hard work if it can be placed on the saw-mill carriage to saw the recesses for the wings, and run a saw-draft outside of each wing, and some inches deep, for wedges. These may be run a little deeper with a handsaw, to allow the wedge to penetrate to the back end of the wings. Great care must be taken in banding and wedging, as it is difficult to fasten them afterwards, should they get loose.

You may also make a pattern from the wheel-draft, for the arms, as it will facilitate and insure accuracy in laying them out. This is merely a thin board of the length and breadth of the arms, with the dovetail of the tenon laid out on the outer edge, at one end, and the slit to admit the rims on the other. To fasten these in the shaft, make two keys with a slight taper, and sufficient size to fill the mortise; place one with the head down, and enter the point of the other between it and the arm, and drive it home. These, as well as the arms, had better be coated with white lead or coal tar.

When large felloes are to be broken out for rims the following plan saves labor: Mark out the felloes by the pattern, and cut the plank to the lengths; then slab and turn down a large log on the saw-mill carriage, run the saw into the middle of it, and take off the feed, fix a bar across between the fender posts the thickness of the plank above the log. Lay a piece of the plank on the log and under the cross-bar, in front of the saw. Start the saw, and feed the plank against it by the mark, which will be easily followed by placing a mark upon the cross-bar above it, and having a man at the other end behind the saw to help. If the circle be small, put in an old narrow saw and set it all to one side, which, with careful attention, will cut a pretty short curve.

CHAPTER VII.

EXPERIMENTS WITH WHEELS.

UPWARDS of twenty years ago I commenced a series of experiments with different kinds of water wheels, and being the owner of a grist and a saw mill, where the quantity of water is rather small, but with a considerable fall, I have continued these experiments at intervals ever since, with the expectation of being able to construct a wheel that would equal the overshot in power, without its complicated and cumbrous machinery.

I began with wheels working on the reaction principle, which were much in vogue at that time. I fixed up a disk, or part of a wheel, a foot in diameter, with an orifice at the centre to admit water from a flume, with ten feet head. This I fastened upon a shaft, which remained the same in all the experiments. I then made another portion or disk to correspond with the first; to this last I attached movable pieces of wood or iron, as the form of the issues made most convenient, and screwed the two together. After trying this with one form of issues and noting the result, I unscrewed the movable part, took out the central pieces forming the issues, and replaced them by others of another form and figure, until every conceivable curve and position was tried, taking care to have the *sum* of the issues of each to be the same, whatever their *numbers* might be. To test the different powers and velocities of these, I laid down an inclined plane, upon which I placed a small stone boat; this was loaded with small stones and had a line attached to it, the other end of which was made to wind around a small shaft of the machine. By varying the load on the boat, and noting the velocity with which it was hauled up the incline, the merits of the various forms were compared and tested.

I mention these experiments thus particularly, only to say that they satisfied me that no experiment with a small model, however accurately made and figured up, for a wheel of working size, and another head of water, can be relied upon; as upon making wheels by the data furnished by these experiments, for driving my 4½ feet millstones, first with 22, and afterwards with 30 feet head, I found the result to differ each time; and had to make another set of experiments to find the best mode of applying the water in practice.

I will not tire the reader with a full detail of these experiments, but merely state that after a trial of upwards of twenty different kinds I adopted a modification of Dr. Barker's wheel, Fig. 24. All these experiments

Fig. 24.

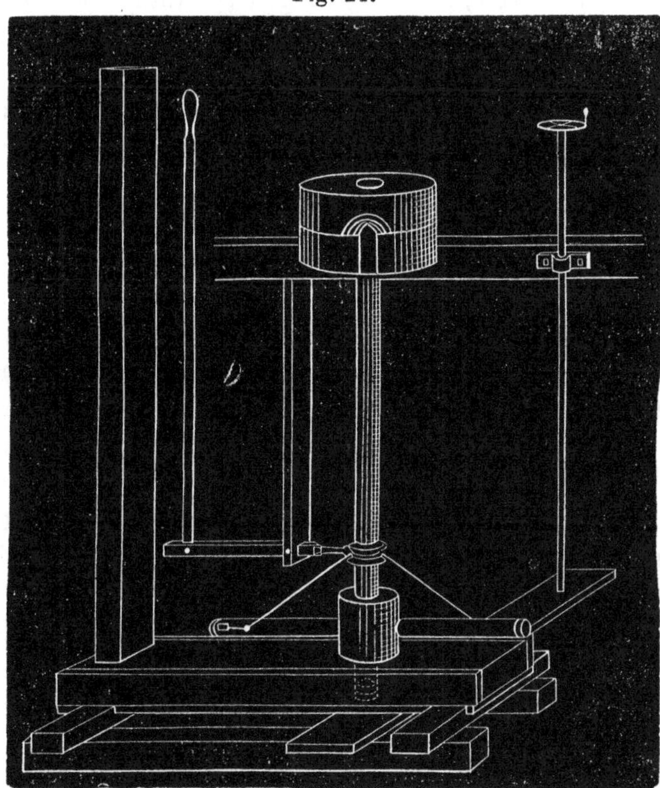

were made by placing the various modifications of wheel on the lower end of the sam. iron shaft or spindle; the millstone was placed upon the top without gearing, except a small pulley upon the spindle, from which a belt drove an upright shaft with bevel gearing to drive the bolt and elevators; and the water was admitted under the bottom. As this is a very simple and effective method of driving millstones, I will describe it in detail.

Barker's Wheel.

Bring the water in a strong tight spout about six inches deep and two feet wide under the wheel. Under and clear of this, place a bridge-tree, thin, wide, and movable at one end, for raising and lowering the spindle and stone; make a round hole a foot, more or less, in diameter, according to the head, to admit the water to the wheel, and a small hole, four or five inches in diameter, in the bottom; turn a piece of maple or other hard wood to fit this small hole; fill it full of tallow and place it in the hole, and resting on the bridge-tree, the upper end being turned to fit the concave end of the spindle, which rests upon this block.

To make the wheel, get a central piece cast, about the size and form of a patent pail, with parallel sides and a strong bottom, with a hole in the centre to admit the shaft, upon which it should be keyed at a proper distance from the end. This must have two holes in opposite sides, about eight inches in diameter, to admit the hollow arms, leaving about three inches of whole metal below them. This part must be turned off true. Make two pieces of plank, with a semicircle in each, to fit the outside of the casting, and cut a recess around the under side of these semicircles about half an inch deep and an inch or more in width; nail around these recesses two or three thicknesses of leather that has been soaked full of tallow, the edges projecting past the wood, so as to come in contact with the turned iron and make a tight, free, joint. These are to be fastened on to the top of the spout with handy screws, and then adjusted.

To make the hollow tubes or arms, get two pieces of thin boiler plate of the requisite length and breadth, say three and a half feet by two feet; bend these round like a stovepipe, and rivet the seam; compress the outer ends a little, as it will allow them to revolve with less resistance from the air; cut an end to fit, either of iron or wood, and rivet the iron tube in over the end to keep it tight. Cut a hole in each arm about two inches from the end, on opposite sides—each hole about 2¼ inches deep and 3 inches long—turn the iron out from the centre so as to form a ledge above and below to direct the discharge, and rivet a like ledge on the outer end of the hole. Now make a slide of the same iron of sufficient length and breadth to cover the hole, with the outer end turned out to form a ledge for the inner end of the hole, and fix an eye or loop in the other end to come out through a slit in the tube behind the hole, and of the same length. This slide is placed inside the tube with the bent ledge or end projecting through the hole, which with the loop projecting through the slit keeps it in place. If you take hold of this loop and draw it back towards the centre the hole will be entirely open, with a ledge all around; push it towards the end half way and the hole will be half open, half shut; push it all the way to the end and it will be shut, the end ledges being in contact. This arrangement admits of the issues being enlarged or diminished at pleasure, while the full pressure of the water is maintained, which is a desideratum not easily obtained with other kinds of reaction wheels.

The following is the method used for commanding

these slides: Place over the shaft a strong double ring, or band, with a groove around the centre of it; take a rod of iron and loop one end into the loop of the slide, the other end into the ring or collar on the shaft; fix both slides the same—the connecting rods being of such length that when the ring or collar is down to the wheel, and the two rods nearly in a line, the two slides will be closed. Raise the ring or collar up the shaft a certain distance, and the slides will be opened. Now fix a lever with a forked end to catch round the centre hollow part of this collar, and with range of movement from the shut to the open position of this collar, and the slides can be moved quite easily while the machine is in motion. I arranged this collar and rods to the slides with the intention of attaching a common ball governor to work them, but as I only used these wheels to drive a millstone the governor was not required.

This whole arrangement gave good satisfaction, and rather more power than any other tried on the reaction principle. But it involves the expense of strong tubes or spouts to conduct the water to it. We may dispense with these by the following arrangement, which, though not so suitable for a grist mill, may do better for some other kinds of machinery:—

Make a round wooden tube of a diameter proportioned to the power proposed, and a length equal to the height of the fall (see Fig. 25). This tube I made by sawing a suitable log down the middle and hollowing out the two halves, and then banding them together with iron; the lower end supported on a concave journal running on a wooden step, and having a strong circular disk instead of wings, the upper end having two cross irons at right angles, in the centre of which the upper journal was fixed. The sheet iron tubes are put into each half before they are put together; the inner end being put through the wood and split into narrow portions, then

Fig. 25.

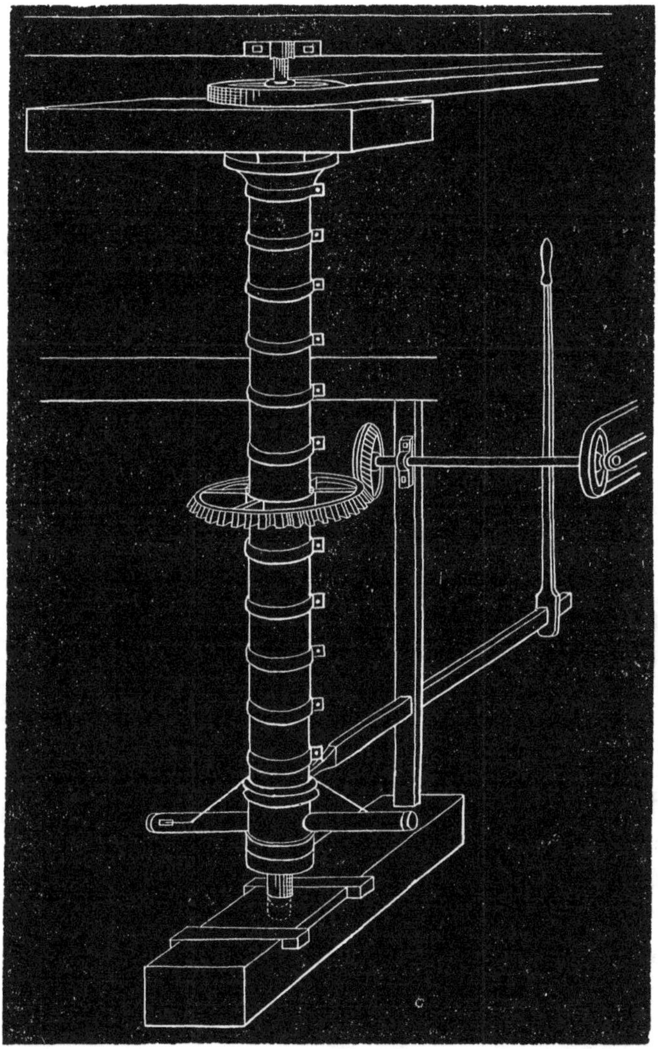

turned back and nailed into the wood to make them fast.

The upper end of this hollow shaft must be widened into a funnel shape to admit the water. Care must be taken to gauge the quantity so that the tube will be full and not run over. The following plan answers well: Make a gate in the bottom of the spout to admit the quantity necessary, and beside this main gate make a small valve gate, hung near the centre; to this valve fix a small stem or rod in such a position that when the gate is shut the stem shall point in the direction the machine and water are to revolve; but when open it shall hang perpendicularly. Attach a wooden float or weight to this stem at the proper height for the water, and the arrangement is complete. Open the main gate to admit as near the quantity as possible, and the valve gate will regulate the rest. When the water falls too low, the weight of the float will hang perpendicular and open the valve; when too high, the float will be carried round and raised by the water, and thus shut the valve. Without this apparatus the water will either get too low and lose part of the head, or too high and run over; because the machine discharges more water the faster it runs, the centrifugal force increasing the pressure of the water towards the ends of the tubes, and thus increasing the velocity of the discharge and adding to the power.

I tried many experiments in expectation of getting more power by running these slower and with shorter arms, and then placing another wheel outside of them, like a light tub-wheel, to get the power of the discharging water, but found it better to increase the diameter and speed of the first wheel and use it at its full speed, than to divide between the two. It made a simple machine very complicated, besides losing a part of the centrifugal force, which in these wheels is a very considerable item, and helps to compensate for friction and the power required to take the water from a state of rest, and impart to it the requisite revolving motion.

This will be better understood if we remember that this machine inverted is the common centrifugal pump. Place the end of the upright tube in the water to be raised, with a valve opening upwards at the lower end, and fill both upright and cross tubes with water through a hole at the top. Plug up the hole and cause the machine to revolve, and the contents will be discharged through the hole at the end of the arms; at the same time the suction or vacuum will open the valve in the bottom, and the water will rush up and supply the place of that discharged. Place a circular trough around to catch the discharging water, and the pump is complete. It will be observed that in this pump, as well as in the wheel, the hollow arms only require to move, the main tube may be stationary, and either perpendicular or in any other direction.

This is an excellent pump where a continual discharge is required, and the requisite motion can be had for it; but as it acts by suction, or rather the pressure of the atmosphere, it will raise water only a little over thirty feet; as a water wheel, however, it is applicable to any height of water, and the higher, the more economically it will work, on account of the *vis inertia* of the water, before alluded to; the only limit being the strength of the confining tubes and other component parts.

When steam is used to propel a rotary machine in this way its power is wasted. Steam is so light, elastic, and ethereal, that every increase of pressure only causes it to escape the faster, without any corresponding reactionary force being imparted to the machine; while water, being heavy and incompressible, requires a great increase of pressure to double or quadruple the discharge; and the power of the wheel being as the pressure per inch multiplied by the number of inches of the area of the discharging orifice, it will be easily comprehended how these wheels answer better with a high head. A head of thirty-three feet being equal to the pressure of the atmosphere, gives a force of fifteen lbs. to the square inch, which being applied at the end of a 4-feet lever, as in driving a millstone, the whole power is easily computed and compared with any other power applied by a pinion or sheave, the lever power of which is half its diameter.

To understand the principle upon which these wheels act, let us suppose the discharging orifices both closed by their slides, and the trunk and tubes filled with water. It is evident that there would be a pressure upon every inch of the inside surface of these tubes equal to the pressure due to that head of water; but the machine would not move. Now open the slides, and the pressure upon so many inches as the hole measures is removed, while the pressure opposite the hole remains the same as before, and the end of the tube will be pushed back with a force equal to the pressure per inch, multiplied by the number of inches in the hole, and it will incline to revolve with a force equal to the sum of both holes thus found; or, if all obstruction be removed, it will revolve with a velocity a little over that due to the head of water, measured from the discharge to the surface in the trunk, as found by the Table. The increased velocity of the wheel over that of the water is due to the extra pressure of the water caused by the accumulated centrifugal force within the arm.

A question in relation to these wheels has arisen in my mind, which is of considerable importance, and though yet unsolved, I will try to explain. It is a fact, well known by experiment, that water issuing through an orifice in a thin plate of metal, like these tubes, discharges a much less quantity in a given time than is assigned by theory; but by fixing a ledge around the orifice, as I did with these wheels, it forms a kind of tube by which the quantity discharged is greatly increased, and by a peculiar form of tube, the discharge and consequent velocity can be more than doubled. The question then is: Whether the *increased* or *diminished* discharge is more economical as applied to these wheels? If the force be simply the reaction or recoil imparted by giving impetus and velocity to that column of water issuing through the orifice, then the greater the velocity, the greater will be the recoil. But if the power depends only on the internal pressure, then the simple orifice will be the best, as that may be enlarged, while the pressure elsewhere will remain the same. This will allow about double the number of inches of orifice, with the same expenditure of water.

I may explain that I discontinued the use of these and the central discharge wheels, next described, and substituted overshots in my mills, on account of a want of sufficient water at all times for the quantity of work to be performed, experience having satisfied me that although either of these kinds of wheels would drive *one* run of stones with as little water as an overshot, when two or more stones were attached to the same overshot, it would do more work with the same water. The reason is, that the small wheel has little else than the spindle and stone to drive, and nearly all the power of the water is applied to grinding; while the overshot requires a considerable quantity of water to drive itself and the intermediate machinery at the proper working velocity; and it is only the quantity applied over and above that amount which drives the millstones and does the grinding. And when the wheel and all the machinery with one run of stones are working at the proper speed and power, a small addition to the quantity of water will enable it to drive another run at the same rate. In other words, it is not necessary to double the quantity of water to do double the work, in the one case; while in the other, every increase in the quantity of work done requires a proportionate increase in the quantity of water expended.

A method patented a few years since for boring pump logs by hollow cylindrical augers of any required diameter, will facilitate the making of this kind of water wheel, as the requisite tubes can be bored and turned of solid wood, and with very little banding; whereas it was quite a job to make them of staves or in two halves, as formerly, and band them sufficiently to prevent leakage.

CHAPTER VIII.

CENTRAL DISCHARGE WHEELS.

A NUMBER of carefully conducted experiments with centre vent wheels of small size and with a low head of water, resulted, like those made with reaction wheel models, in no practical benefit. The reaction wheel which gave the best result in the model had only one issue, coiled around in a spiral, or rather helical, form—like the main spring of a watch—from the centre to the circumference. Expecting the same result in practice, I made a wheel seven and a half feet in diameter, with a similar issue of considerable size commencing at the centre, and gradually contracting as it wound around towards the circumference, where the water escaped through an opening of three by four inches. This channel was about thirty feet long, and the water issuing through it under thirty feet head had the appearance of soapsuds, and instead of accumulating power and velocity, as the model indicated, the friction through the long helical tube destroyed both.

This experiment, although in itself a failure, furnished a useful practical hint concerning the central discharge wheel and its scroll case, namely, that as it is impelled around the inside of the scroll by the head pressure, the water has a constant tendency to continue on in a straight line, which forces it against the outside of the water way with a much greater force than it exerts at the inside; and this constant force and friction destroy a large proportion of its power. Besides, it is only the inside or lesser pressure, and not the outside increased force, that can be applied to any wheel placed inside of this water way. To test the difference between the pressure of the water against the outside and inside of this water way, I bored several auger holes around both the outside and inside of the scroll in pairs opposite to each other, and at equal distances apart; a piece of lead pipe was inserted in each of the first pair of holes, and a full head of water was let on, and the height to which each tube forced up the water was noted. The tubes were then inserted in the next pair of holes, and the first plugged up: thus it was continued, and the difference noted all around the scroll. The result was, that the tubes next the outside of the scroll threw the water one-half higher than those next the inside; and the height of all the jets gradually diminished as they came round to complete the circuit. Other similar experiments showed that the larger the circumference of the scroll, the greater was the waste; and, also, that the waste and loss increased in proportion to the increase of the head and velocity of the water.

These experiments tend to explain what we have a hundred times remarked in practice, namely, that these scroll-cased central discharge wheels answer well for a low head and ample supply of water; and that when the head and velocity are high, and the quantity limited, they give a power much below that ascribed to them by theory.

After many and varied experiments of this kind, we concluded that in order to make one of these wheels that could compete with an overshot, two conditions were necessary: First, to get a throat or adjutage through which the water would issue with the full velocity ascribed to it by theory; and second, to apply all this acquired force upon the wheel. The first can be accomplished by issuing the water through a short tube made conical or funnel-shaped at the inner end, in the form called "the contracted vein," elsewhere described; and the second condition is obtained by connecting this tube or chute with the wheel in such a manner that the bucket becomes a part of the chute, or takes the place of it, receiving the water at its full velocity, and by the gradual curvature of the bucket, combined with the centrifugal force produced by the motion of the wheel, gradually and completely overcoming and absorbing the power of the water.

This will be better understood by examining the cut, Fig. 26, where it will be seen that the bucket receives the water nearly straight at its outer end, and by its curvature and the motion of the wheel it brings the water around in a circular scroll towards the centre,

Fig. 26.

where, its force being spent, it drops down clear of the wheel by its weight. By this arrangement, it will be seen that the curved bucket virtually takes the place of the outside scroll of the common centre vent wheel, and receives the whole force of the water, intensified by the curvature it here undergoes. The force of the water is still further augmented and impelled against the bucket by the two opposite forces which here act upon it, that is, its acquired impetus, which urges it onward in a straight line, and the centrifugal force, which urges it outward from the centre. These contending forces increase the pressure of the water against the inside of the buckets by counteracting its progress towards the centre of the wheel, and gradually overcome and arrest its impetus altogether, and it falls clear of the wheel. It may be further observed that the friction of the water against the buckets and inside of the wheel as it passes through, which in all reaction, and most central vent wheels, seriously retards its motion, is, in this, a help instead of a hindrance, as it acts in unison with the motion of the wheel.

This is an excellent wheel for a high head of water, being more simple in its construction than Barker's, and not, like it, impeded by the *vis inertia* of the water, nor by friction, and, when properly calculated and made, gives a result fully equal to the overshot. To obtain this result, it is necessary that the size and velocity of the wheel should be nicely adjusted to the velocity of the water; so that in entering, the water may move enough faster than the wheel to enable it to overcome the centrifugal force, and follow the inside curve of the bucket to its inner end, where the water should arrive with its momentum exhausted.

To obtain these results we tried the following experiments: We fixed a wooden disk or rim upon the lower end of the iron spindle of a millstone, the disk being four feet in diameter, and another rim of the same outside size, but only twelve inches wide, being open at the centre to allow the water to escape. This last rim was under the other, the wrought iron buckets being between. The form and position of sixteen buckets were then calculated as nearly as possible by making a draft upon the floor, and by this we made a pattern of the space between two buckets, and cut brackets out of half-inch stuff by this pattern to put in between the buckets upon both rims. These were nailed in all around, and a small half-round bolt put through at the back of each bucket to hold the rims together. We tried this, but found it too large in diameter to give the 4½ feet stones the requisite speed, 160 revolutions per minute, with 30 feet head. We turned off a portion of the rims and repeated the experiment several times, each time making corresponding alterations in the brackets, and position and shape of the buckets, until the wheel was reduced to 3 feet and 1 inch, and the lower rim to 6 or 7 inches wide, when we were satisfied with its working, and made patterns and got wheels cast, with the wrought-iron buckets between, requiring no bolting.

We had seen many central discharge wheels in operation, and patterns, drawings, and descriptions of many more, and in every case the water was conducted to them through a scroll-shaped water way extending around the wheel, although some had several gates for the admission of water; and we determined to try how one of these wheels would work in such a scroll. We accordingly made a very nice scroll, the outside of smooth iron, and the top and bottom lined with zinc, and delivered the water through this upon the wheel, the throat being a little over 3½ inches square, or 12 inches of water under 30 feet head. This would hardly give the requisite velocity to the stone when running empty and free, and could only be made to grind between one and two bushels an hour by checking it up quite slow. We then removed a portion of the scroll at the end, and brought the termination close to the wheel to shoot in all the water, and found upon trial that it worked better; another portion was then removed, which showed a corresponding improvement. Thus we continued to cut and try, until the scroll was reduced to the small portion we had previously been using, which was only sufficient to shoot all the water into three buckets out of the sixteen, and then its former power and velocity were restored, and it would grind eight bushels of wheat or corn per hour, or average one hundred bushels of mixed grain or custom work per day of twelve hours.

There will be some waste by this short scroll, unless it be continued the true circle of the wheel, to cover the space between two buckets, otherwise a portion of the water will escape past the point of the chute without entering the buckets. In fact, the curve of the waterway, and form and motion of the wheel, cause some water to pass in this way in spite of every precaution against it.

Fig. 27.

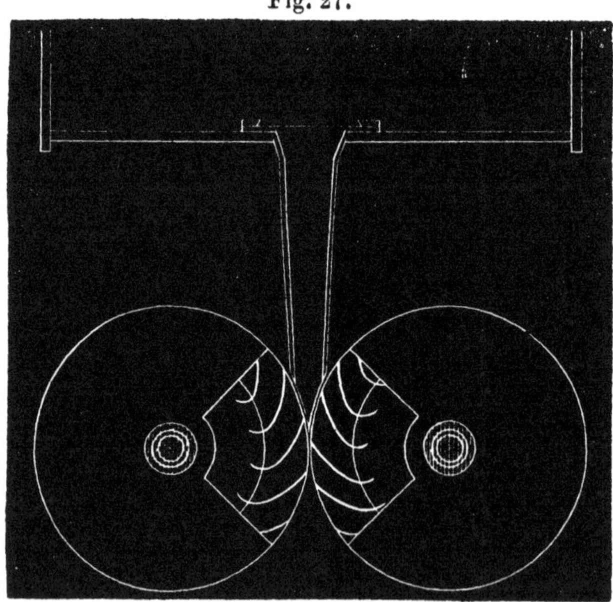

To obviate this waste and loss we constructed a double wheel, Fig. 27, the water being shot in and divided between the two, as shown in Fig. 27. These we made thirty inches in diameter, with a corresponding alteration in the form of the buckets, although under the same thirty feet head. The wheels were connected by cogs around their circumference, both above and below, both being cast from the same pattern, but turned with different sides up, to bring the buckets right. The buckets terminated in corresponding cogs, so that when placed in gear they overlapped each other, thus preventing the escape of any water between the two wheels. These buckets stepping in between each other at equal distances, receive the whole direct force of the water from the short directing issue without any waste or loss by friction, and split the stream equally between them, each carrying its particular portion of water around with the motion of the wheel until its force is expended, when it drops down and escapes beneath. It is essential that these wheels should be geared permanently together, to keep the step of the buckets, as if allowed to shift their position with regard to each other, the overlapping ends would come in contact and damage each other. The connection, in some cases, can be made more conveniently between the shafts above; and where a horizontal motion is taken from these upright shafts, the best way is to place a shaft across the top, with a double bevel pinion in the centre between the two, gearing into a corresponding bevel pinion on each water wheel shaft, as shown in Fig. 28.

These wheels work equally well on a horizontal shaft. The short chute of the single wheel should commence at the centre of the descending side and terminate at the bottom. We have placed several of these wheels in this position on the crank shafts of saw-mills, both for English gate and gang. The wheels being cast with a double set of buckets, thus discharging the water from both sides; that is, the wheel has a central rim which extends to the shaft, where it bulges out equally on both sides into a strong hub, by which it is keyed on to

Fig. 28.

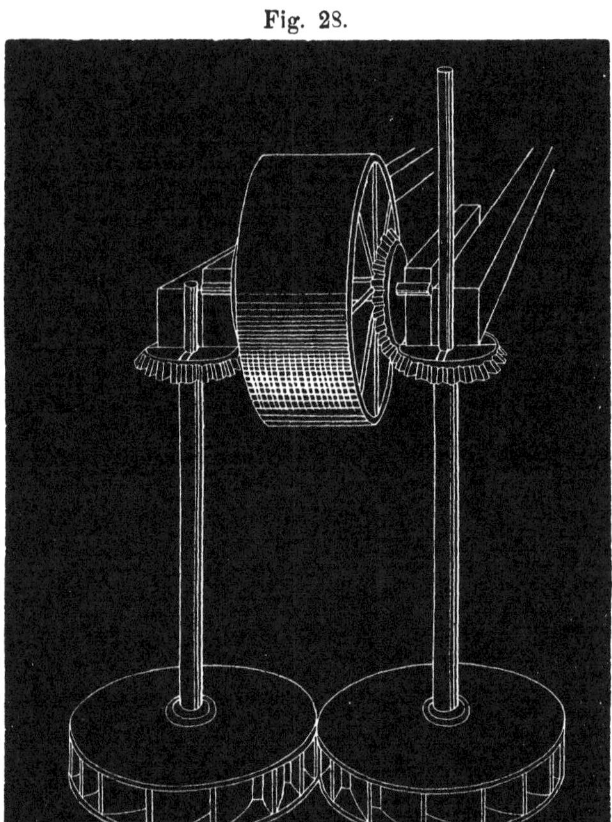

the iron shaft. Buckets made of thin boiler plate are cast into each side of this rim, those on one side breaking joints midway between the others, to avoid weakening the rim, and an open rim only five inches wide is cast on to the outer edge of both sets of buckets. The water discharges through these, the inner ends of the buckets being slightly curved outward, and the bulge of the central hub assisting to throw the water out endways.

The top or entrance to the short chute is covered by a planed iron gate, water-tight, which is slidden off or on by a turned stem working through a stuffing-box. Either the single or double wheel can be supplied by bringing the water down a straight tube; or a cheap and good plan is to bring it from above in what is called a rainbow chute, being the natural curve which the water would assume if allowed to pitch over and fall without interruption, and only confined sufficiently to direct its shape and protect it from disturbance by the air. This chute is only for a horizontal shaft, and if properly made is supposed to give a velocity at the wheel equal to that which any heavy body would acquire in falling through a space equal to the height of the fall, except a slight reduction for the resistance of the air, and friction against the directing chute.

This is the velocity ascribed to the water by theory, and is the standard by which its velocity is measured and computed for any particular head, but, as before remarked, the *actual* velocity, found by experiment, differs widely from this, according to the nature and form of the throat, or adjutage. Thus an orifice through a thin board, or plate of iron, will admit a far less quantity in a given time than this velocity would indicate, while by using a short cylindrical tube, of the same bore as the orifice, it will discharge as much water in three minutes as the simple orifice will in four; and if the length of the directing tube be only about twice its diameter the increase will be still greater.

By examining a jet of water issuing through a thin plate, it will be seen to taper and diminish in size from the plate to the distance of about half the diameter of the hole, where the area of this smallest part will be to the area of the hole as 10 to 15 or 16. This contraction is what is called "the contracted vein," and by enlarging the end of the tube next the flume in these proportions, to a funnel shape, the same contraction takes place before entering the tube, and the flow is consequently greatly increased; by adding an enlarged conical piece to the outer end of this, thus making the tube conical at both ends, the velocity of the discharge can be more than doubled, being to that of the thin plate as 23 to 10. Care must be taken not to overdo this, as the length of the outer conical piece should not exceed three or four times the diameter of the small end, and the sides should not diverge more than to form an angle of three or four degrees with the other. This is a subject with which every millwright should endeavor to make himself thoroughly acquainted, as well as with the retarding effects of friction in all the various forms of chutes or adjutages, and the influence of curved or scroll-shaped water-ways upon the velocity of the water.

By thoroughly understanding this disposition of water in motion, and forming the short directing chute to accommodate and take advantage of these peculiarities, the water can be brought in contact with the wheel at the full theoretical velocity; and by proportioning the size and motion of the wheel and curve of buckets to this velocity, according to the head, the whole velocity and force of the water can be arrested and absorbed by the wheel, without waste or loss by friction. The attainment of these two conditions is all that is required to solve the long pending problem "How to make the most of a fall of water," the only other desideratum being, that the wheel be such as will admit of an easy and cheap connection with the machinery to be driven. This the wheel under consideration admits of, at a tithe of the expense required to gear the overshot to machinery under the most favorable circumstances, the cost of the wheel itself bearing about the same proportion to that of the overshot (*i. e.*, one-tenth).

When intended for a wooden shaft both rims are made open in the centre, the buckets projecting past the rims to the inside about two inches. Four cross-arms are mortised through the shaft, and the wheel is secured to them by screw bolts.

We have recently made several patterns for wheels and chutes of this kind, of different sizes, and had the wheels made and set to work since the chapters on the overshot and reaction wheels, in the foregoing part of this work, were written. This will account for any discrepancies between the above remarks and those made in former chapters, as the wheel and application of the water are new, so far as we are aware, and we have not yet had time to calculate and reduce the velocities and curves to a rule or standard whereby they may be theoretically computed.

There are many other varieties of central discharge wheels, each kind being, like our own, recommended by its inventor and patrons as the best. Most of these are, as we have already said, supplied by a scroll water-way extending clear around the wheel, generally a single gate to admit and control the water: one kind that is common in the Northern and Western States having three gates, each having a scroll or water-way encircling one-third of the circumference, and terminating close to the wheel. This is a good wheel for a low head and plentiful supply of water. Another that has come into extensive use within a few years is called Leffel's Turbine, Fig. 29, and has twelve gates, and as many separate water-ways; this wheel the makers claim to equal an over-shot with any fall or quantity of water. We have never seen this tested, but confess that we cannot see how a

Fig. 29.

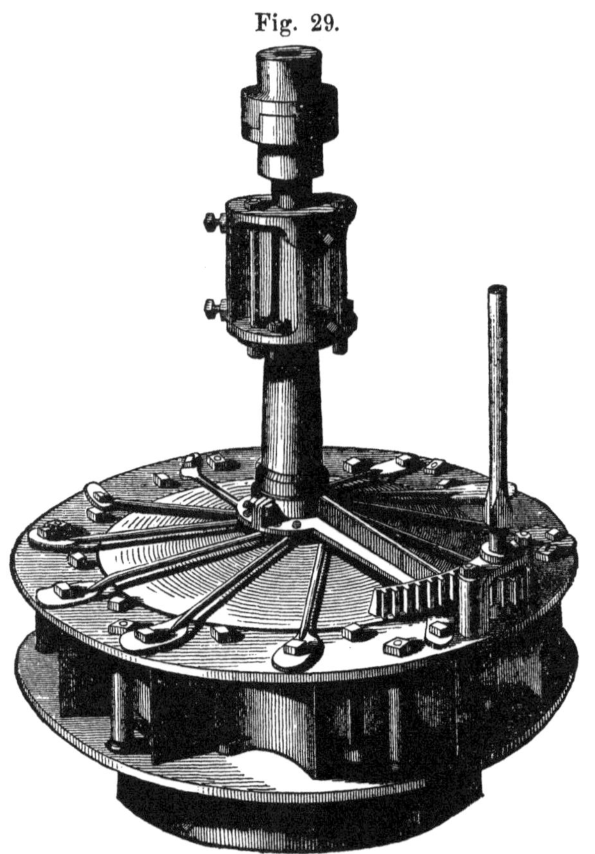

small quantity of water with a great velocity, divided into twelve separate streams, each impeded by a certain amount of friction against its gate and directing chute, can be a very economical application. On the other hand, each of these chutes applies the water straight and direct upon the opposing bucket, without the loss and friction of conveying it around the whole circumference in a scroll; and this peculiarity, with the excellent workmanship and material used in their construction, accounts for the high reputation they have gained wherever they are used.*

If we were urged to give a candid opinion, we would say that the Leffel wheel ought to give as good a result with a low head, and up to a moderate height, as any other turbine with which we are acquainted, not excepting our own; but with a high head and velocity, we would back our own against it with perfect confidence. We are confirmed in this opinion by another consideration to which we have not before alluded, viz., that the principle upon which the Leffel wheel is constructed involves a diminution of size, and consequent increase of velocity, proportioned to the increase of head of water. This with a high head reduces the size of the wheel to 10, 8, or even 6¾ inches in diameter, and increases its revolutions to twelve or sixteen hundred per minute, which involves a reduction of speed by intermediate machinery to enable it to drive mill-stones or other heavy work. This, of course, is called *gaining power*, being the reverse of the intervening machinery of the overshot, for *gaining speed*, but both really waste power, and the less machinery interposed between the wheel and the work, the more will the power of the water be transmitted to the work. The fact of these small wheels running at such a velocity for years is a proof of their superior workmanship.

Other central discharge wheels have perhaps twenty or thirty of these chutes or water-ways. The direct action wheel of Howd, and others similar to it, are examples of this kind. They answer well on large and

* These wheels are manufactured by Poole & Hunt, Baltimore.

sluggish streams which afford a plentiful supply of water, with but little head, as by enlarging the size of these wheels almost any amount of water may be used to make up for the deficiency of fall; but they are not suitable for a high head and small quantity of water.

There are many other central discharging wheels with only one water-way or chute, which is made in a scroll form, and of more or less capacity, diameter, and depth, according to the head of water under which they are to act, and the amount of power to be developed. Some of these, like the wooden centre vent and the Tyler wheels, work by the impulse of the water. Others, as Riche's wheel, the Ring Wheel, and Little Giant, work partly by impulse on the outer incline of the bucket, and partly by reaction through the contracted inner issues. There are probably not less than forty different varieties of these wheels in use in this country, differing from each other only in some slight particular, and all working on the same general principle of intercepting the impulse of the water by the outer circumference of the buckets, and delivering it at the interior, with the impulse or velocity more or less appropriated and applied by the buckets. And the *degree* of *perfection* with which the velocity is thus absorbed and applied, is the measure of excellence of any particular wheel.

The wheels last mentioned are much used in grist mills; there is generally a wheel for each run of stones, with the shaft geared to the spindle, either by cog-wheels or a belt, although the gearing is frequently dispensed with, and the stone placed directly on top, and the wheel on the foot of the same iron shaft or spindle. These wheels answer well with a moderate head and supply of water, but should not be used where the quantity is limited and head high, as in that situation they involve too much friction and waste. Neither should they be used where the head is very low and the quantity ample, because the principle does not admit of applying sufficient water in such a situation, and the Jonval, the Leffel, the Howd, or similar wheel with numerous chutes, which may be adapted to almost any quantity of water, will answer the purpose better. When the conditions above referred to are favorable, these are a cheap, durable, and convenient kind of wheel, although not well adapted to run on a horizontal shaft.

When a millwright has decided to use this description of wheel, he may in general select the particular kind or name, according to his fancy, merely adapting the capacity of the wheel to the particular site and work. In doing this, regard must be had to the nature of the stream; some being comparatively clean and free from floating obstructions, while others contain a variety of rubbish which would choke the issues of some kinds of these wheels, and be a continual source of annoyance, and others, having a larger throat, would swallow and dispose of everything that could get through an ordinary rack.

We have seen several Little Giants, which is an *excellent* wheel on a clean stream, condemned and taken out because they would choke up with bark and leaves, and even coarse sawdust, and require to be cleaned out even ten times a day. These wheels were protected by a rack so fine that it had to be cleaned much oftener than the wheels; and if this was neglected, the spaces became so completely closed as sometimes to allow the water to draw down in the flume, until the pressure upon the rack crushed it completely in. This occurred on the Trout River, in the State of New York. But another mill on Norton Creek, using the reaction wheel on the Fourneyron principle, with narrow issues, was periodically clogged with eels every fall, at nearly the same time. The dam was built across the outlet of a marsh twelve miles long, and pretty wide, raising the water about two feet over that whole area. The eels took a fancy for migrating down stream from this pond every year, at the commencement of winter. They glided through the *rack* without any difficulty, but the result shows that they had not studied the nature of fluids *theoretically*, and their practical experiments had never been extended as far as *hydronamics*, for they always blundered and got entangled in the wheels, and these would have to be stopped and cleaned out several times in the course of a night.

When we used Barker's wheel we were frequently annoyed by muskrats getting into the hollow arms and closing the issues in a similar manner; these had to be chiselled out, as no provision was made for getting out such obstructions, and we depended on a fine rack for keeping them out, but the muskrats would either climb over, or force through by bending the rods of which the rack was composed, and their strength, as shown under such circumstances, was astonishing. A convenient scuttle should always be placed in such wheels, for these reasons.

When these scroll-cased centre-vent wheels are used for driving saw-mills, they are generally placed upon an upright shaft, from which a horizontal shaft is driven by bevel gearing, and the motion is taken from a drum on this shaft to the pulley on the saw-arbor, or crank-shaft, by a belt. Several saw-mills with Yankee gangs and large circular saws are in operation on a small stream, a branch of the Trout River, before mentioned. The upper one, Tucker's mill, with gang, large circular, and clapboard machine, is driven with Little Giant wheels, with about 24 feet head; it is indifferently attended, and does but little business. Goodwin's, half a mile below, contains about the same amount of machinery, with some 26 feet head, and is driven by Reynolds' turbines; these are probably as good wheels as any of the class, and the mill being well attended does a good business. The next is Rood's mill, a large circular and English gate, driven by a Ring wheel, and a pair of Hildreth's, with 13 feet head; this mill does a good business for the head of water. Jordan's mill, with Yankee gang, English gate, shingle and lath machines, and edger; this has 18 feet head, and is driven by the author's wheels, these being on the crank shafts without gearing: the chute admitting water to the gang wheel, taking 120 inches, that to the English gate 60 inches, and to the wheel on the vertical shaft, which drives all the other machinery having two chutes, one of forty-two and the other forty-eight inches area, in sections, either or both of which are used as the machinery being driven is varied. The head over these gates and chutes is only about fourteen feet, which makes this a much smaller quantity of water in proportion to the head, than any other mill on the stream. Below this is Hare's mill, a gang and English gate, driven by Rose wheels, a pair on each crank shaft, with about ten feet head; and Judge Adams' mill, a large circular, clap-board, and shingle machines; these are driven by wooden centre-vent scroll wheels, that driving the circular having an area of chute of over four hundred inches, with about eleven feet head.

The Ring wheel, driving Rood's circular, with thirteen feet head, measures $20 \times 23 = 460$ inches in the throat of chute taken the same way. The others mentioned, we had no opportunity of measuring; and the reports as to the quantity of water admitted were so conflicting, that they were not reliable—those given we measured.

The Jonval wheel,* Fig. 30, although it discharges at the centre, like those we have been considering, works upon a different principle from them, and is rather the type of the Spiral Discharge Wheels next described.

Fig. 30.

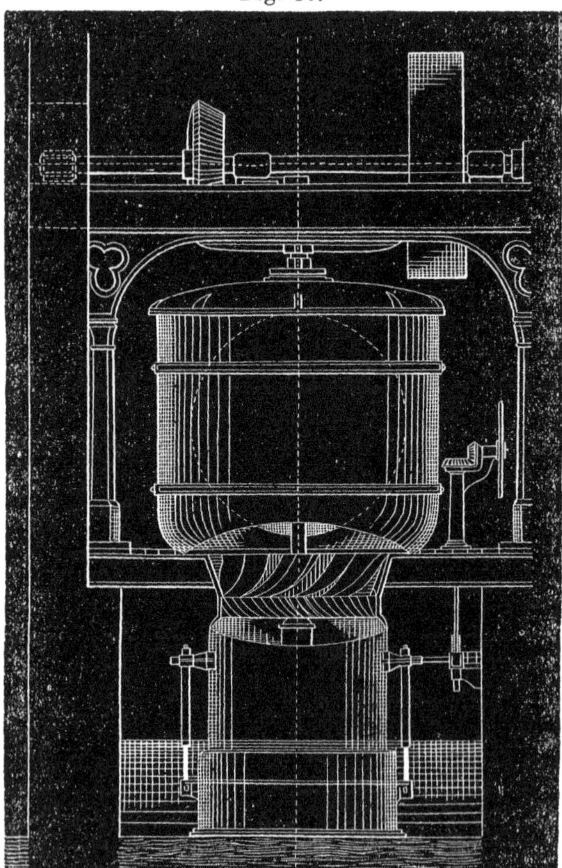

It is probably the best wheel in use for obtaining a

* Manufactured by Emil Geyelin, 15 S. Seventh Street, Phila.

great and *constant* power from a large quantity of water, either with a moderately high or low head. But where the power requires to be varied, or where the quantity is limited and the fall very high, it has some defects. The Leffel wheel is better adapted to develop a *variable* power, economically from the water, than the Jonval. And the author's double centre-vent wheel, which applies the water in one straight unbroken chute, will yield more power from a small quantity of water, with a very high head, than either the Jonval or the Leffel wheel. Under the last conditions, Barker's reaction water wheel will also beat them. For experiments on this subject see the article on "Barker's wheel."

The principle of the Jonval wheel applies the water upon the whole area, from the central axis to the circumference, and thus admits of the greatest possible quantity, while the spiral guide plates, by starting the water in the direction of the wheel's motion, where the oblique buckets intercept it and reverse its direction and discharge, give it at the same time the utmost velocity obtainable from the water. This peculiar combination makes it probably the very best wheel in use for a low head and constant quantity, but a great waste is the consequence of diminishing the full power, by checking the flow through the gate; because the water must still fill the guide scrolls and issues of the wheel, and the power is thus only diminished by transferring a portion of the force and velocity of the water from the wheel to the gate beneath it. The great expense of this wheel is another objectionable feature which precludes its use on a small scale, and confines it mostly to large establishments. The Jonval, like the Leffel and other crack wheels, is so well advertised and known, that any further notice of it or them here is quite unnecessary.

CHAPTER IX.

SPIRAL DISCHARGE WHEELS.

THERE are a great many varieties of spiral discharge wheels in use, generally in pairs, upon horizontal shafts. In this way they are well adapted for saw-mills, especially where the head of water is low, as the oblique position of the floats or issues, tends to screw up the velocity with which they revolve: like the vanes or sails of a common wind-mill; thus giving sufficient velocity to the first mover, and dispensing with intermediate gearing. To understand this increase of velocity, observe a boy's toy wind-mill in a moderate wind; you will see that it revolves with a velocity far greater than the breeze that impels it. The reason is, that the vanes are held and revolve in the same plane, with their oblique face exposed to the same constant force, instead of flying away from the wind or receding before it, as in the case of the various wind-wheels on vertical shafts, and others that are driven direct before the wind. In this case the velocity of the sail must be deducted from that of the wind, and the remainder is all that can possibly apply to the sail.

This is the reason why a vessel will sail faster with a side wind than with it fair in the rear. When the vessel is moving fast before the wind, nearly as fast as the wind wants to follow, there can be but little force exerted on the sails; but when the wind is coming in a direction oblique to that of the ship, the sails are constantly pressed upon and filled, being held to the wind by the keel and rudder, as the sails of the ordinary windmill are held by the shaft. This side pressure upon the vessel, and end pressure upon the shaft of the windwheel is much greater than the actual propelling force, and this is also the case with the spiral discharge waterwheel, and greatly increases the friction where a single wheel is used upon a shaft. But when a pair of wheels are placed upon the same shaft, facing each other, and supplied with water between, then one balances the other, and in this way they work well upon a horizontal shaft, the water discharging towards each end, clear of the wheels. But this arrangement, with a vertical shaft, is not so economical, because a space must be allowed beneath the lower wheel for discharge, then a considerable space between the two, for a free supply of water for both, and then the other wheel inverted above that, and discharging upward.

All this causes a great waste in a low head of water, and these wheels should not be used with a high head. I have seen several wheels working in this way, when the upper wheel was a hindrance rather than a help, as the lower wheel had so much more pressure that it threw the water out through the upper one, which acted as a brake, besides diminishing the pressure upon the lower one. In general, a better result can be obtained by removing the upper wheel, and replacing it by a smooth "lighter" of the same size.

A better way to place a single wheel of this kind is to inclose the water-way clear to the shaft; as this reduces the space for friction and waste of water to the circumference of the shaft, instead of that of the outer edge of the wheel. In every case the water-way should lead in towards the side in a scroll form, so as to give the entering water a revolving motion in the same direction the wheel is to move.

The simplest and perhaps the best way to place such a wheel upon a vertical shaft, is to place it in the bottom of the flume, and case it around as you would the bed mill-stone in the floor. This insures the full and free force of the water, and also prevents trouble from frost, but where several wheels are used in the same flume, each must have its compartment partitioned off, and its gate, or some other arrangement for putting on and off the water. This I have done by two small hatchway doors, shutting one each side of the shaft, with half size of shaft cut out of each. The first one is hard to open, and hard upon the hinges. I have also hung this by four slings, to two rollers across above the flume, so as to swing back; in this way one door is sufficient, if you make a slit in it of the width of the shaft and from the forward end to the centre. Swing this forward, and it will cover the wheel, except the slit that straddles the shaft. This must be filled by a stationary piece, which is little interruption to the water, if the cover is placed some inches above the wheel. This arrangement works easily and is not liable to get out of order. A separate compartment of flume for each wheel is the best where it can conveniently be obtained.

The downward pressure of the water being added to the weight of the wheel shaft, &c., in this arrangement, makes a great friction upon the foot gudgeon, which must be large in size, and turned out concave in the end, and run upon a block of hard wood fried full of tallow. Great care must be taken to admit the water to it freely, as otherwise it will burn out in a few minutes, even when quite deep in back water.

There is such a great variety of these wheels in use,

that it would be needless to try to enumerate them, and useless to specify the slight modifications got up by the different makers, and insisted upon by them and their advocates, as being of the first importance, and the last improvement. Equally useless would it be to detail the controversies of these interested sticklers about angles and curves; as to what angle the issue or float should form with the plane of the wheel—whether said issue should be an inclined plane or a curve—and what curve—and whether the issues should pass directly through in a line with the axis, or with a draft towards the circumference, &c. &c.

It may be remarked, however, that the more oblique the issues are, the nearer the wheel will run to the full velocity of the water; and that, by giving the issues a draft outward, like the reaction wheel, it increases the inclined working area, without diminishing the internal pressure or increasing the expense of water; and also secures a slight advantage from the centrifugal force, and a freer clearance from the discharging water.

In conclusion, we would advise those intending to use such wheels, to select from those that have been tried and approved, rather than incur the expense of novel and untried improvements; taking care to get them of a suitable diameter and area of discharge for the work intended; and remembering that they all give a discharge greater when in motion than when at rest; while many of the central discharge wheels give less. This fact must be borne in mind, in comparing or testing the different kinds of wheels, with the view of replacing one kind by another; as some use more, and some less, than the actual measurement of water. And lastly, remember that these wheels are well suited to give a speedy, direct motion, with a low head, and ample quantity of water, but are not suitable for a high head, with only a limited supply.

The Rose Wheel.

A very good modification of the spiral discharge wheels last described, is much used in the Northern States, and is called "The Rose Wheel." This wheel is made with open issues or buckets, passing obliquely through it, like the former, but instead of all the issues around its whole circumference being exposed to the action of the water at once, like that wheel, the Rose wheel is driven by a spout, or water-way, approaching it in a line tangent to the wheel's circumference. This spout or "boot," as it is called, is nearly straight, being only curved at the end sufficiently to cause the water to strike the two or three opposing buckets at right angles, the peculiar curve of the buckets turning the water in the opposite direction, as it passes through them.

This wheel is equally well adapted either for a high or low head of water, and either on a horizontal or vertical shaft. It can be used either singly, or in pairs, although a *pair* of these upon an upright shaft would be subject to the same inconveniences as the spiral discharge wheels in a like arrangement. That is, a space under, for discharge, and a space between, for entrance of water; then the up-hill manner in which the upper wheel discharges its water. These, together, are sufficient to condemn that plan; besides, a much better and cheaper way is to put another equal discharge of water upon the opposite side of the same single wheel.

This double chute I have used upon my central discharge wheels, already described, the principle of which is similar to these. It gives double power, with but a very small addition of either expense or friction, and either one or both chutes can be used, as circumstances may require. Both the Rose wheel and central discharge wheel last mentioned answer well upon a crank shaft for a saw-mill, although the double chute cannot be used so economically with a horizontal shaft, as the curved point of the chute in this case ought to terminate at the lowest part of the wheel; and it is better to put on a pair of Rose wheels, facing towards each other, with a double boot between.

CHAPTER X.

SPIRAL, OR SCREW FLOOD WHEELS.

THESE wheels are very seldom used or seen in this country, and, in fact, they are but little known, and the principle upon which they act but little understood or appreciated, except as it is adapted to the spiral discharge wheel, last described. Their principle of action is the same as the screw propeller, which has in a measure superseded the paddle wheel in steamboats—the difference being that the propeller is driven round (revolved, by the steam-engine, and its oblique vanes forced against the dead water behind, thus pushing forward itself, and everything connected with it, while the screw water wheel is located and remains in its place, and is driven or revolved by the force of the passing current against its oblique vanes, and transmits this motion through the gearing to any machinery attached to it.

To comprehend this similarity better, take a screw propeller, and place its axis upon suitable bearings, and parallel with the stream in a strong uninterrupted current, and entirely submerged, and it will furnish a motive-power to drive machinery—the amount of power being in proportion to the area of the vanes, and the force and velocity of the passing current. These remarks are only intended to illustrate the similarity of the two principles, and not to recommend the propeller as a suitable flood wheel, because the action of each being exactly contrary to the other, like the steam-engine and force-pump, requires a modification in structure and details, to fit either and each to its particular purpose.

We have put in several of these wheels to drive light machinery sometimes in comparatively small rivers, and never failed to get the full power calculated and required. As little could be learned from the experience of others with regard to these wheels, we had recourse to experiment, and found that much more power could be obtained by making separate detached vanes or wings, thinned off at the edges—like the blade of an Indian paddle—than by placing a close continuous thread of a screw around the shaft the entire length. The reason of this is, that the first or up-stream end of the solid screw breaks the force and spreads the current, so that little or nothing is gained by extending the screw and shaft further down stream: while detached vanes, like those referred to, allow the water to pass on each side, which keeps up a continuous flow of current clear to the shaft. This current, though partially checked by the vanes, is restored by mingling with the uninterrupted current all around, and ready to act with full force upon other detached vanes, placed further down on the shaft; and by extending the shaft, and a judicious distribution of these vanes, or paddles, along its whole length, a portion of the force and velocity may be abstracted from a large volume of water, and applied to any useful purpose, giving a power sufficient for a grist or a saw-mill, or any ordinary machinery. The whole-thread screw checks only a volume equal to its own area, and does not get the full benefit of that, owing to the dead water behind the vanes, against which they work. For some further reference to these wheels, see the chapter on the Undershot Wheel.

When such wheels are made to work in a rapid current, the vanes may be set more obliquely in the direction of the stream; but for a current comparatively slow, the ends should point nearly across the stream. In all cases they should be " weathered" in regular proportion as they approach the shaft (centre); that is, a kind of mould-board twist—like the vanes or sails of a wind-mill—which gives to the inner end, moving slowly in a small circle, considerable obliquity, and diminishes it gradually towards the outer end, which moves swiftly and in a large circle; and here the face of the vane, or sail, should stand nearly in the direction of the plane of its motion. This subject is referred to in the chapter on the Spiral Discharge Wheel, and the subject of weathering the vanes being fully discussed and explained in the chapter on Wind-mills, need not be repeated here.

The motion is sometimes taken from these wheels to the machinery by an endless chain, composed of alternate open and close links, graduated to the same equal step, and working around and connecting two clutch wheels, one upon the wheel in the water, the other upon the machinery to be driven. These have projecting clutches or cogs around their circumference, at equal distances, corresponding with the open links of the chain, which step into these to prevent it slipping, like the chains in a tread-horse power, or the feed chain in a gang saw-mill. Light machinery may be driven by an ordinary short link cable-chain, by turning out a groove in both wheel and pulley for it to work in. If the chain be short, or nearly perpendicular, it must be furnished with a tightener, because it stretches fast, especially when first used. There are so many links (joints), that a little wear in each is multiplied considerably in the whole length. If the chain be of some length, and nearly horizontal, it will not require a tightener. We have frequently driven four and a half feet burr mill-stones with such a chain. It is not equal to a belt, for swift, light motion, but answers well for a slow, heavy motion, and can be used in the water, where a belt is useless.

Such chain gearing, for the purpose under consideration, answers well in warm weather, but is an everlasting source of trouble in frosty weather, as it is continually carrying up water sufficient to cover with ice everything within the reach of its influence. For this reason it is better to place a narrow-rimmed cog-wheel around the down-stream end, and take the motion from this by a pinion and shaft. This is more reliable at any time than the chain, and gives no trouble with the ice, the lower end of the shaft and pinion being—like the water wheel itself—wholly under water; the only part exposed to ice is at the surface of the water. The whole structure must be guarded by a boom, or breakwater, placed obliquely up-stream above it, to shoot floating ice or flood-wood past. If a gear wheel of large size be placed on the water wheel, its arms should be placed at an angle and weathered, like the vanes or paddles; it will then help as much as hinder the revolutions of the wheel.

Mills of this kind, driven by the current of a river,

have been common in some parts of Holland for several generations. These Dutch mills are sometimes built to float upon the water, and can be moved at pleasure to any other suitable locality. They are anchored in a rapid of the stream, or made fast to the shore, and provided with a gangway to communicate with the land; when custom begins to fail in one place, they remove to another; and thus possess a decided advantage over mills built upon *terra firma.*

The first machinery that we ever saw driven by a wheel of this kind, was a grist mill of three run of stones. It was built upon the Genesee River, perhaps midway between the city of Rochester and the Alleghanies. It was before the Genesee Valley Canal was made, and boats passed up and down the river past the mill. The mill was built with one end on the bank, which was perhaps twenty-four feet high; here the door communicated with the road; the other end projected over the river, and was supported upon naked posts. In the rapid under this projecting end of the mill the flood-wheel was placed. A bevel pinion, without arms, was placed upon the down stream end of its wooden shaft, and another wooden shaft, with a corresponding pinion gearing into the first one, stood up perpendicularly from this end, both shafts having their bearings in the same wooden block; this perpendicular shaft reached up to the mill floor, and had a large spur wheel at the proper distance from the upper end, around which the three run of stones were placed, and by which they were driven.

The shaft of the water-wheel was eight or nine feet long, and fifteen inches in diameter. The screw was of one continuous thread, standing out something over two feet from the shaft, and was composed of pieces two and a half or three inches thick; each piece being narrow at the end next to the shaft, and spreading out wide at the outer end, to close and complete the circuit. They appeared to have been split out of a winding or twisted tree, which gave nearly the required weathering or mould-board shape, with little dressing, and the narrow end of each piece was tenoned and morticed into the shaft. A wide bar of iron was bent and fitted to the proper shape, and spiked securely around the outer edge of this continuous screw, thus fastening the outer ends of all the pieces composing it together, and making one piece help another. This wheel was inclosed in a plank box, open at both ends and top, except when a large gate closing the whole up-stream end, was shut down to stop it. This stopped the current through the box, and the wheel stood still; when the gate was raised, the current was renewed, and the wheel started.

When we were at this mill, it had been idle three weeks, on account of back water; the waters were abated, and it had just started. It was situated near the head of the Genesee Flats—these extend along the river for twenty miles, and average four miles in width—when the spring freshet comes down from the Alleghanies, it first fills the river channel, which is from twenty to thirty feet deep, and then floods the whole flats to the depth of several feet; this causes dead water at the bottom of the channel, and accounts for stopping the wheel, and thus explains the mystery.

CHAPTER XI.

MILL DAMS.

A VERY important appendage to almost every mill is the dam, and the failure of the dam from defective construction, or other causes, has spoiled more mill speculations, and ruined more mill owners, than any other defect.

It is in many situations a very expensive item, and in some it is exceedingly difficult to make a dam that will stand; there are such differences in situations, and such a great diversity of surroundings to be taken into account, that it is impossible to give any general rule or direction that will apply in every case. The situation and circumstances must determine, in a great measure, the materials of which it should be constructed: whether of logs or stones, of timber and plank, or earth, or of part or all of these combined; and the solving of the problem, how, and of what materials to construct a dam, to be most economical and safe, is often the most severe task and test of the experience and ability of a millwright. Still it is quite common for a man of good sense and judgment in other matters, to imagine that little skill or experience is necessary to construct a dam, and with the assistance of a few laboring men, he will build his dam, at the same time that he employs experienced millwrights to do the rest of the work. But it is needless to say much on this subject, as some will see and acknowledge the truth of these remarks at once, while others will only be taught by the expensive school of experience.

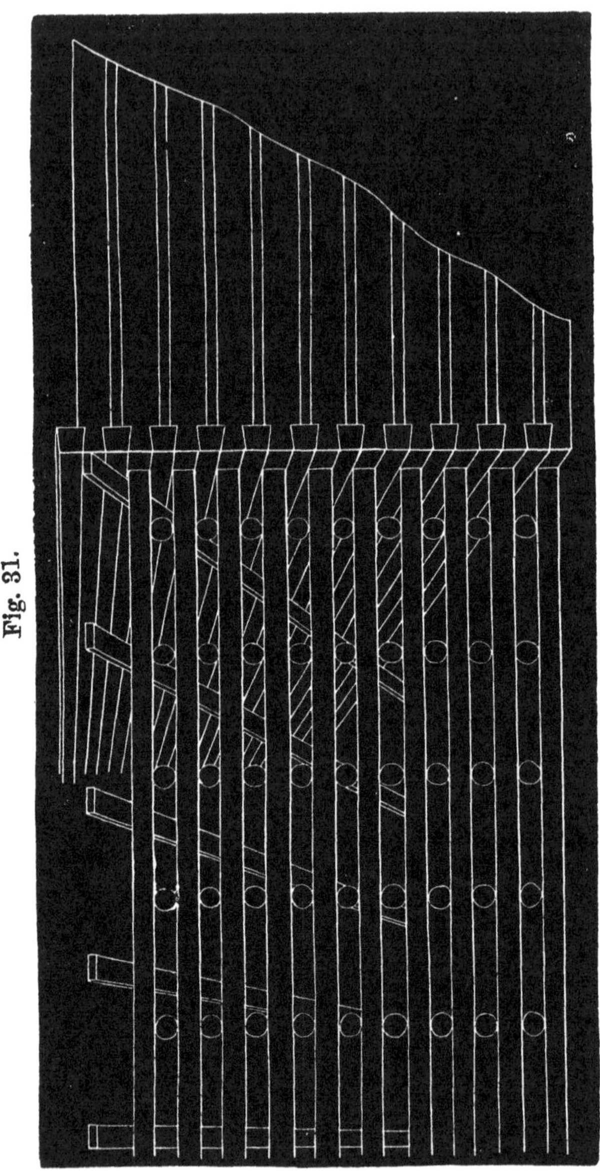

Fig. 31.

We will proceed to describe several different plans for constructing dams with different materials, and begin with LOG DAMS, Fig. 31. These, in a locality where timber is plenty, are cheapest, and easiest to build. If the bottom be rock or other good foundation, begin by laying a large log across the stream, at the down-stream face of the intended dam, this you will extend from bank to bank, by laying one log at the end of another, having each piece as long and large as possible, taking care to clear away everything that will wash out from under, and where hollow places occur, put short logs across under, so as to give it a safe foundation. Then put short logs across this six or eight feet apart, their butt ends lying upon the log and their top ends upon the ground, up stream from this; you will now place another tier upon these, above and parallel with the first one, but inclining slightly up stream; then another set of short ones, their butts upon the last tier, and top end upon the ground beside the first cross ties. These must be a little shorter than the first ties, to admit of laying a smallish log on the ends of the first ones, and up into the angle formed by the second ones; you can now lay "skids" upon these small logs, and proceed to roll up your third tier of large logs, along the faces. Care must be taken to notch them a little where they cross each other, to insure their laying safely, or block them secure with a stone or piece of wood where the small ends come.

Your next tier of ties must be notched well down at the small or up-stream end, and you must proportion your two parallel tiers of logs and these ties, so that the front or breast will rise enough faster than the rear, to keep it like a portion of an arch, and have the cob-work, when finished, fit the rafters; that is, the larger tier of logs at the breast should support the rafters near the top, while the smaller tier at the rear, should support them near the middle, and the lower ends of the rafters rest upon the rock or bottom. It will be seen that a breastwork, so constructed, is like a portion of an arch or circle, of which the foot of the rafter is the centre, and the front of the breastwork the circumference; and the more weight is put upon it the stronger and more solid it becomes. Care must be taken not to carry it too high, or steep, for the length of rafter (or radius), as in that case the force of water behind might slide it away in a body.

If logs are convenient, this may be covered with them, like rafters, touching each other, taking care to fit them well and chink the cracks. The moss on trees and old logs, in damp places, is good to chink these cracks, as it grows and increases in such a place, instead of washing out. Cedar bark, pounded soft like oakum, is also good. Such a covering requires but little gravelling to make it tight, as the pressure of the water forces the packing down into the seams formed by the round logs, where it is not easy to wash it out, or displace it by any other means.

Such a dam is cheap, strong, and durable, where there is a constant supply of water; but on small streams liable to dry up in summer, and allow the logs to dry and heat and check, they very soon rot, and are therefore not to be recommended for such a situation.

Where logs are not so easily obtained for covering, place rafters about three feet apart, upon the cob-work, with the butt ends down, and plank crossways upon these as in Fig. 32, using thicker planks near the bottom, if the dam is deep, while thinner ones will answer near the top, except where the water is intended to run over, where strength is required to withstand the ice, flood wood, &c. The lower planks should be scribed down carefully to the bottom, and all the others bevelled about one-eighth of an inch off the upper or inside corners, and halfway through to allow dirt to get in to make it tight; the ends should be sawn open inside

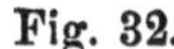

Fig. 32.

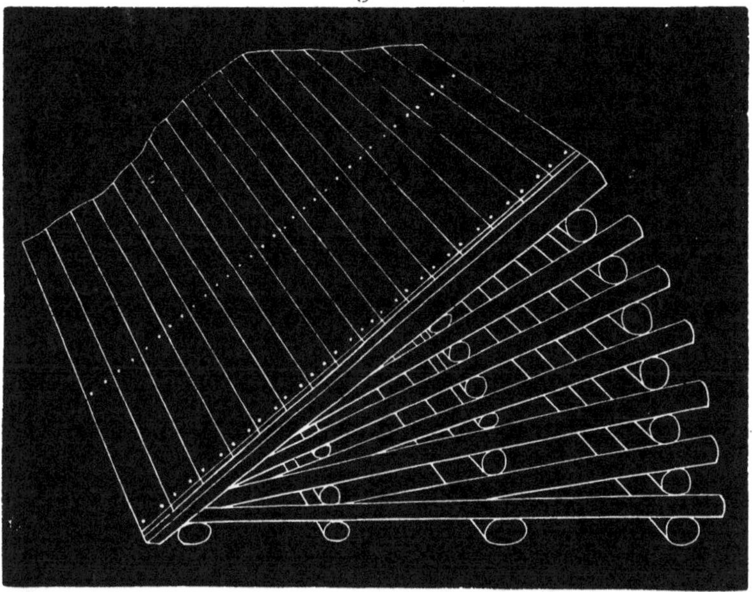

also. If the plank were partly seasoned, and put on with pins, each pin being draw-bored a little, to pinch the plank tight, then a little gravel along the bottom, and a little sawdust stirred in when the water is first raised, will make it tight. Pinning and draw-boring is a tedious process, however, and it is more common now to spike the plank on and cover with gravel.

Great care must be taken with this, as with every other kind of dam, to splice it securely on to the bank, at each end, and one principal consideration should be to attach it in such a way, that, as the dam settles down (as it certainly will, to some extent), it may settle *on to* the bank, and secure it as much as possible against frost, and not *away from* it, to cause a crack. These ends, and the place where the waste water runs over, are the most difficult parts to secure.

So far, we have supposed the dam to be built upon a rock, or other good bottom, and in such a situation there is little trouble with the waste water; but upon a soft bottom it is a very different affair. A carefully constructed chute and apron are then indispensable; and in some bottoms, composed of sand or mud, it is necessary to drive piles to a great depth, and across the whole breadth of the river, to get a foundation sufficient for building a dam. In one instance a double tier of these piles was driven by machinery, the first twenty-five feet long, and the second row breaking joints upon these, fifteen feet long; these were driven into the quicksand with the lower ends slightly inclined up stream, and the tops bolted to a stick of timber. Upon this the dam was built, composed of crib-work of timber filled with stone at the centre, each end being flanked by a wing of frame-work, rafters and plank. The whole channel, for some distance above, was filled up with "fascines," and the banks below were protected by the same means.

These fascines are small bundles of brushwood, tied up by three withes, about a foot apart. They are placed in courses like shingles, each course overlapping another, the upper end inclining down stream, and each bundle fastened down by three stakes. These stakes are three or four feet long, the small end sharpened and a pin hole bored through the butt end, in which a pin about eight inches long is driven. These stakes are thrust through the fascines, one at each withe, and in an oblique direction, being entered above the withe, and passing through below it at the under side. They are driven down by wooden mauls, until the pin comes upon the withe, and hugs the whole bundle down tight.

Another dam, in a similar situation, was afterwards built, without the pile work, the bottom and edges being protected by fascines alone; this dam has now stood for many years, and appears to be safe and durable. And yet another was made by rolling in loose stones, large and small, just as they came, mixed plentifully with pea-straw, until the channel was filled completely up, and formed a rapid. This last stands equally as well as the others, except that a large tree imbedded in the bank, which was not discovered at the time, caused a small break the first season.

These dams are all upon large streams emptying into Lake St. Peter, in Canada, near the sandy north shore of which lake they are constructed. It is worthy of remark, that all these streams carry a large amount of sand and mud in solution in their waters, especially during freshets, which very soon fills all the interstices in these fascines, or pea-straw and stones, and renders the whole mass solid and compact. This result would not take place in a stream where the waters were clear and free from sediment, and the filling would have to be supplied as the work progressed.

A very good way to make a foundation for a dam upon a soft bottom where timber is plentiful, is to lay a tier of long logs lengthways of the stream, and touching each other all the way across, and build the breastwork of the dam upon the up-stream end of this causeway, allowing the rafters and covering to run down past the end, to prevent the water from getting under it. The down-stream end forms the apron for the waste water to strike over upon. Care should be taken to fit the foundation logs well down, as they will not accommodate themselves to inequalities in the bottom, like the stones and straw, and they should be placed as low as possible to insure their being always wet, and thus protected from decay.

In a locality where stone is plentiful, a stone dam may be the most suitable. There are several ways of constructing these. Where the foundation is good they may be built nearly perpendicularly on the face, an incline of two inches to the foot being sufficient. The back part may be carried up with considerable incline, making the wall quite thick at the bottom, and then filled up with earth behind, so as to make the surface considerably flat. In this case a bond timber should be built into the wall several feet from the top, on the up-stream side, and another along the top, and the space planked between these, at least as far as the water is intended to run over; otherwise the force of the water will carry the earth over near the top, and displace the stonework. The best and safest way is to place rafters upon the stonework, and plank upon these, the same as with a log or frame dam.

The following is an excellent method of constructing a stone dam, where the bottom is not fit for an upright face, like the last mentioned. It requires a good deal of material and labor, with little other expense. Commence by laying a tier of large flat stones across the stream, below the point intended for the brink of the dam, then fill to the level of this tier for a considerable distance up stream, with gravel or stones of any size or shape, anything in fact that will not wash or waste away. Now lay another tier of flat stones across the stream, as before, on the top of the first tier, but receding up stream; say six inches from the edge of the first, if the stones are thin, and more if the courses are thick, calculating for each course according to the thickness, to bring the declivity equal; fill up level on the up-stream side, by dumping in as before, and it is ready for the third course. Thus proceed, step by step up stream, until the dam is of the required height, when a slight coat of good earth will complete it. The water may run over such a dam the whole length, and is, therefore not liable to wash the earth over, as it will where it must all run over a narrow apron. Such a dam, when completed, and the water running over, has the appearance of a natural fall over a ledge of rock, being so thick at the bottom, and the water flowing down the incline, broken by the steps or ledges of the different courses, makes it the safest and most durable dam that can be made. Another very important consideration in favor of this kind of dam is, that the dam is complete and safe at any height; it may be built only a few courses high the first season, and when time and circumstances are favorable, it may be increased again and again, until it is completed; while almost any other kind of dam requires the expense and labor for its full completion to be laid out at once, before it can be used, or even tried with safety.

Any stone dam is stronger and safer to be built with a slight bow or arch up stream, the ends being well set into the banks, which answer for abutments. This form directs the overflowing water towards the centre of the channel, thus saving the banks from washing, and in a dam like that last described, where the water is flowing over the whole length, it has a very fine appearance.

Any kind of dam is stronger and safer to be curved a little up stream, but when built of wood, either frame or logs, or cribwork, this form is more difficult to construct than a straight line; for this reason, as well as its being the shortest, the latter is generally preferred.

There are several different ways of constructing

frame dams. Where the bottom is good and level, frames may be made in a triangular form like a drag, or new land harrow. These are placed with the narrow end up stream, three or four feet apart, and in a line from bank to bank, and planked up along the upper angle. If the bottom is good these may be placed upon it without any further preparation; but if soft or uneven, a tier of timbers must be laid down across the whole distance, one under the point, and another under the wide or down-stream end of these triangular frames for a foundation, as in Fig. 33.

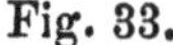

Fig. 33.

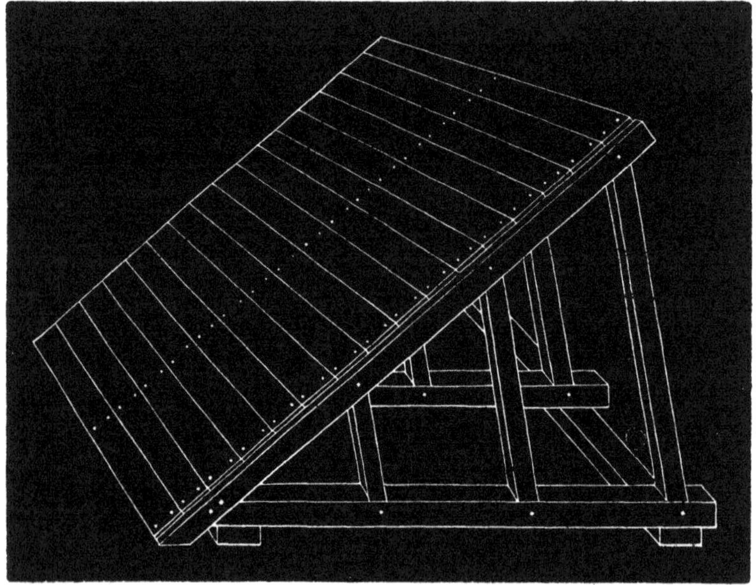

In situations where these foundation timbers are found to be necessary, the following plan will be more economical. Place one tier of timber across the stream at the breast, or down-stream side, another tier at the up-stream side or foot of the rafters, and a third tier in the middle between these. Frame posts into the down-stream and middle tier of timbers, to support another tier above and parallel with these, the posts slightly inclined up stream; the first tier of posts being sufficiently long to raise the timber they support, to the intended height for the top of the rafters. The posts to raise the centre timber to be framed of proper height to range with this last and the bed piece under the foot of the rafters. This arrangement gives three supports for the rafters, at the foot, top, and centre; and by notching the rafters slightly down upon these points, it binds the whole framework together sufficiently strong when the planking is fastened on, and the weight of water comes upon the whole, to make it safe and secure. This is probably the cheapest and easiest way of making a frame dam, as the timber requires very little preparation, and the work of framing is comparatively small.

A strong and substantial dam may be made of crib-work, filled with stones. This is made by placing two tiers of timber parallel with each other, at a proper distance apart, and dovetailing ties across between them at certain distances to bind them securely together, and then filling the space with stones. The sides are built up with some incline, and it is planked on the up-stream face to make it water-tight, beginning at one end and running the plank down endways to the bottom, and spiking them on to the cribwork.

This dam can be pushed across a stream where the water cannot be got rid of, or is too deep to build other kinds with advantage. A portion of the crib can be built in the water at its place, or on the water at a more convenient place, and floated to its destination, and sunk with stones. We have seen such a dam built across a large river, above an old dam, and in the dead water of the pond. The cribs must be open at the bottom, and the stones being tumbled in loosely, settle down into the inequalities of the bottom, and keep it secure. Even if the foundation should wash out in parts under these cribs, the loose stones will settle down into the cavity and fill it up, and all that is required is to fill the crib again with stones to the surface. This dam may be planked over the top, and is then safe with any quantity of water flowing over it.

In some situations, and under certain circumstances, an embankment of earth is the best and cheapest dam. We have built such dams where the embankment was almost completed by ploughing and scraping, requiring only a little earth wheeled or carted on to finish it. Care must be taken to provide a good waste or flood-gate, to allow the water to pass off freely during the progress of construction. A neglect of this precaution caused the writer much loss and trouble with his first dam of this kind. We turned the water out of the channel above the place, and carried it past in a ditch along the bank, and then placed a tube at the lowest part, under the embankment, to carry off the leakage, intending to have a dam perfectly tight; as it was a small stream, although subject to sudden floods. Twice, during the process of construction, a heavy rain caused such a flood that a large portion of the earth was swept away, and it was with much difficulty that we succeeded, the third time, in getting the waste wier completed for the safe overflow of the water, when the third freshet occurred. Since that, we have always placed a flood-gate the first thing, in commencing such a dam; by placing the bulk-head frame and gates on an incline of 45 degrees, or even less, it may be gravelled and made perfectly tight, and is of great use after the dam is finished, in relieving it during a freshet, or in emptying it for repairs, &c.

Great care and some experience are required in the construction of the waste wier for such a dam, as the frost, by heaving up the earth at each end, is almost certain to raise and displace the timbers and planking of the wier, and allow the water to work under. The best method of securing such a wier, or similar connection between earth and woodwork, is the following: Dig a trench across the wier, canal, or whatever it may be, deep enough to be below the frost, and a safe distance into the bank; place a stick of timber in this trench, and spike a tier of plank on it, of such length as will reach nearly to the surface. Place a similar stick across at the surface, over the top of these planks, and place another tier of plank flat against the first, breaking joints with them, and spike this last tier to the upper stick. The first tier of plank being fastened at their lower ends to the stick at the bottom, and below the frost, will remain there, while the other tier being fastened at their upper ends to the stick at the surface, will rise with that stick when the frost heaves up the earth and ice; but as the two tiers of plank overlap each other, the surface, timber, and plank attached to it may be disturbed, while no crack or leak can occur so long as the lower tier remains intact, and the lap is maintained. This will be found the simplest and safest way to connect a spout or flume with a ditch or canal, in any situation that is liable to be disturbed by frost.

An earth dam or embankment of this kind should be raised a considerable height above the highest limits of the water; and it will add still more to its security to cover it with straw, boughs, or brush, to encourage the snow to lie upon it in the winter.

We have seen dams made of cribs of timber, similar to those already described, but instead of being filled with stones, and planked on the up-stream side, the compartments were filled with earth. These are not always reliable; the water is very apt to find a way along the cross-ties, or under them, as these prevent the earth from settling and packing equally; besides, this plan furnishes a favorite resort for muskrats, as these mischievous animals work more under such cribs than in a bank composed entirely of earth. Where stones cannot be had to fill such cribs, and the stream is too large to depend on an embankment of earth, the following plan will make a safer and better dam: Clear and level the bottom as well as possible for a distance of about twenty-five feet, and from bank to bank, and settle down three tiers of timbers, their surface level with the bottóm, and about ten feet apart. If the bottom be soft mud or sand, the centre tier of timbers should be piled, by driving down planks as deep as they can be driven, along the up-stream side; these planks should be grooved and tongued, as they will keep the line, and drive better; they must also be sharpened like a cold chisel, at the lower end, and the advanced corner bevelled off about one-third of the width of each piece. This bevel, with the help of a handspike stuck into the ground and crowded against the outer edge, will keep the tongue in the groove, and insure each piece driving in line and in its proper place or position. If the planks are narrow and not very long, they can be driven by a heavy maul, or otherwise a light tripod or derrick that can be easily moved along as the work advances, may be fixed with a pulley or other contrivance to work a ram. The piles must be spiked to the sill.

A range of posts the height of the intended dam, must

be framed into the middle and up-stream tiers of timbers, if the stream be small and the dam not very high, but if more than eight or nine feet, or a very large stream, posts should be framed into the third course of timbers also. These posts require a plate along the top, and the lower tier must be well braced from the down-stream side. The framework thus constructed must be planked on the *inside*, and the space between the two filled with earth; earth may also be filled in along the up-stream side, next the pond, until it approaches the surface, and slanting back into the pond at an angle of 45°. The space from the middle to the down-stream tier of sills must be planked, for an apron for the water to fall over upon from the waste wier and waste gates. This apron, where the bottom is soft, must be further extended down stream, to protect the foundation from washing. For a high dam, or large stream, the third or down stream mudsills should have posts framed and planked like the other two, thus making a second compartment to be filled with earth; and in this case other sills must be placed further down stream, for the apron and support for bracing.

A great deal of care has been taken to ram down and pack the earth in filling in a mill dam, but this is all labor in vain, as the water will saturate and soften the whole mass like porridge, and in this condition it will settle and pack more equally if the water can be run down once or twice after the earth is thoroughly wet. Unless this is done no earth dam can be depended on for any considerable time.

Probably the most substantial and durable dam that can be made, is that composed of large blocks of stone, cut and fitted, with close joints, and these composed of water-proof cement. Such a dam, as already intimated, is made stronger by being made circular, the top of the arch up stream, and the seams of the different strata or courses, should incline or be lowest up stream; that is, these seams should run back at a right angle from the face, or inclined front of the wall, instead of horizontally, as in ordinary walls. This inclined bearing of the courses is necessary for stability, for the same reason that the face of the wall is inclined, viz: the pressure being all from the up-stream side.

Such a dam requires great care in clearing and starting the foundation, and splicing it into the banks. and also in protecting the ends from injury from the effects of ice and frost; as such stonework conducts the frost to a much greater depth than wood. The expense of building such a dam, even under the most favorable circumstances, is too great to admit of its general use for ordinary mills, and restricts it to large establishments in cities and villages, where the value and amount of business will warrant so much outlay for a dam.

Piers of some kind are often, although not always, required at the ends of a dam, to make a safe splice to the bank, or to protect the bank from washing away in case of great freshets or floods. These are generally made of timber, in the form of cribs, nearly perpendicular on the face next the stream or main dam, and on the up-stream side. These two faces are planked water tight; the other sides are of a shape to correspond with and splice to the bank, whether that be of rock or earth, the crib being strongly built and bound together with cross-ties, and the whole interior filled with earth or stones.

Such piers are only required, when from the nature of the bank, a tight substantial joining cannot be made between it and the main dam; as when the bank is composed of a seamy or shelving rock, or a loose, friable sand, or other soil. Of course the situation must determine the size and form of such a pier, but the height must always be considerably above the highest water known or expected. This precaution is rendered necessary by the excessively high freshets which are known to occur in almost every stream in this country, as the forests around their sources become cleared off, and the land along their courses levelled and smoothed down by cultivation.

While the forest remains in its primitive condition, it not only prevents, by its shade, the snow and rain from melting and evaporation by the sun and wind, but it also causes the earth to absorb and retain the moisture like a great sponge. The leaves, which cover the ground like a thick carpet, commence the process of absorption, which the soil completes by the help of the trees. To understand this, we have only to observe the action of these trees during a storm; as the top and trunk are swayed to and fro by the wind, the roots and soil as far as these roots penetrate are heaved up and let down alternately, until, if the tree is not overturned, tearing up the roots and earth with it, the soil is at least loosened and rendered sufficiently porous to drink up the rain and melted snow, which is gradually let off by springs or otherwise, into the swamps and streams. The heights and hollows made by turned-up roots of trees also tend to prevent the water from flowing freely off the surface of the forest; while these are all demolished by the plough and harrow, and furrows and ditches, and every facility is given the water to run off as speedily as possible.

The clearing of the forests and the cultivation of the soil are the true causes of the many old mills to be seen in some parts of the country, idle and in ruins, that were formerly busy with life and the hum of machinery. The old channels are nearly or quite dry at midsummer and midwinter, and are filled to overflowing during freshets; and for the same reason people now find it necessary upon some rivers, to build their bridges higher, their dams lower, and their booms and breakwaters stronger than formerly.

These booms and breakwaters are necessary to direct and control the logs and timber, and also in some cases the ice and floodwood. It requires much experience and good judgment to locate these properly, but after their plan and position are decided upon, there is little skill required in their construction. The breakwaters are generally piers or cribs of timber filled with stone, and of sufficient strength to withstand the current, with the ice and floodwood, and also the strain upon the booms which are fastened to them. The booms are made in various ways, according to circumstances: when merely intended to stop bodies floating upon the surface, a single stick is sufficient; when intended to intercept an undercurrent, then several sticks, one above the other, are required. A good and simple way to make a surface boom is to make it of single and double sticks alternately, choosing the largest sticks for the single ones. To connect them together, bore two large auger holes through each end, about six inches apart, and mortise the piece between, taking care to have the pairs of equal length. Place the sticks with the end of the single one in the middle, the others on each side, with the mortises all in line, and drive a good hard wood slat through the three, with a pin through each end. Where the boom is intended to rise and fall by the ebb and flow of the water, the pins should not pass through the timbers, but only through the slats, so as to allow the splices to yield, and the boom will accommodate itself to the surface of the water or ice, the fastenings at the ends being allowed a little play for that purpose. The sticks composing such a boom may be squared, or left round; they ought at least to be flattened on the upper side, for convenience of walking upon them, and it will add considerably to their safety to spike a tier of plank along the top of the timbers, the whole length.

Another appendage to the dam is the apron or slide. This is always necessary where logs and timber have to be run over, but upon a rock bottom, where no timber is to be run over, the apron may be omitted. Upon all bottoms of earth or loose stones, some kind of apron must be interposed to break the force of the waste water falling over, otherwise it will eventually undermine the foundation and destroy the stability of the dam. When only required to protect the bottom and foundation, it will be best to make it flat upon the bottom, by placing long timbers across the stream, and covering these by cross timbers or strong planks pinned or spiked on, lengthwise of the stream. Unless this last covering runs back under the dam, it will be necessary to extend it up along the breast of the dam, at least far enough to prevent the water from dashing back under, and disturbing the foundation. On a small stream, not subject to great floods, this apron need not be the whole width of the channel, but on large streams that are subject to great freshets, it is best to have the waste wier as wide as possible, as the extra width allows the increased quantity of water to pass over with less perpendicular rise.

Where a chute or slide is required to facilitate the passage of timber, it ought not to be made wider than is required to pass the cribs safely, as it is desirable to have a good depth of water, and the entrance to the chute must be lower than the rest of the dam; of course the narrower that entrance is, the greater will be the depth of the water in it, and this lower entrance can be raised to the level of the rest of the dam at low water, by movable flash boards, to be removed when a flood is expected. The slide may be a single inclined plane if the dam be not very high, but on very high dams, two or more of such inclines, with a fall or jog between, are better; as these falls tend to break the force of the water, and thus diminish the velocity of the passing timber, and let it over more safely.

These slides require to be strongly made of good timber, well pinned or bolted, and upon substantial cobwork, well secured with cross-ties. If the foundation is rock with a decline or fall below, these cribs may be sufficient without filling; but upon a loose foundation, or one subject to back water, the foundation might be undermined, or a whirlpool or eddy formed by such a chute entering the deep back water, might lift the whole fabric, and carry it away entirely. (This I have seen.) In such a situation, the cribwork must be completely filled with stone, which will protect it against either of these casualties.

The movable flash boards mentioned above, are a valuable addition to almost every dam, especially where there is any deficiency of water. The main dam is made considerably below the high-water mark, to allow the surplus water to pass during a freshet, then by placing pins or other supports at proper distances apart, along the brink of the dam, a tier of plank of any suitable width may be placed on edge against the supports, and made tight; this will raise the water over the whole surface of the pond the depth of these planks. The extra head thus acquired is a good help when water is scarce; besides the extra quantity retained, for we must remember that an inch or two on the *surface* of a pond often contains as much water as a foot near the *bottom*. Whenever high water is expected, these flash boards can be removed and laid aside until again wanted; the supports may be left in some situations, in others they require to be removed. Where much timber, logs, or ice pass over a dam, a good piece of timber should be fitted and well fastened along the whole brink of the waste wier, and also across every point of the slide where the timber, &c., breaks over; these sticks take the wear and tear, and when worn out can be replaced by new ones, and thus protect the permanent timbers of the dam.

In some cases the water continues low and stands so near the same level during the severe winter weather, that the ice gets firmly attached to the whole surface of the dam, and if a sudden rise of water should occur under such circumstances, it would raise the ice, and most likely a portion of the surface of the dam with it. To prevent such an occurrence, the ice should be cut loose from the dam occasionally, and kept clear, especially when a thaw is anticipated. When the ice becomes very thick and strong it will sometimes crack or otherwise injure a dam by contracting during a sudden and severe cold term, and for this reason alone it should be occasionally cut loose during the winter, as it will then give way at the weak line, where it has been cut, and save the dam from injury.

CHAPTER XII.

SAW-MILLS.

The English Gate.

THE business known as the "Lumber Business" is so extensive and varied, and its ramifications penetrate and extend throughout all the subdivisions of this vast continent to such a degree, that few, if any, of the other branches of trade can equal it in magnitude. But with all the different divisions of this vast trade we have little or nothing to do, excepting the simple process of sawing the different kinds of logs into the various grades of lumber. The almost infinite subdivisions which these grades have now assumed, from heavy timbers and planks, down to laths, shingles, veneers, and even matches, with the innumerable variety of machines which human ingenuity has devised to facilitate the manufacture of all these, would fill a volume as large as this is intended to be, though merely giving the very shortest intelligible description of them all, which would be a feat that no single individual could ever hope to accomplish. We will therefore only attempt to describe a few of the most important, and begin with the common "English Gate."

When nothing more than a single gate is intended to be put in, the frame is generally about forty feet long, by twenty feet wide; the lower story from eight to eighteen feet high, according to the height of the banks, and the head of water. Two sills are first placed in position, with five posts in each, the two up stream, and the two down stream pairs having beams framed across between them, and well braced; the centre pair has no

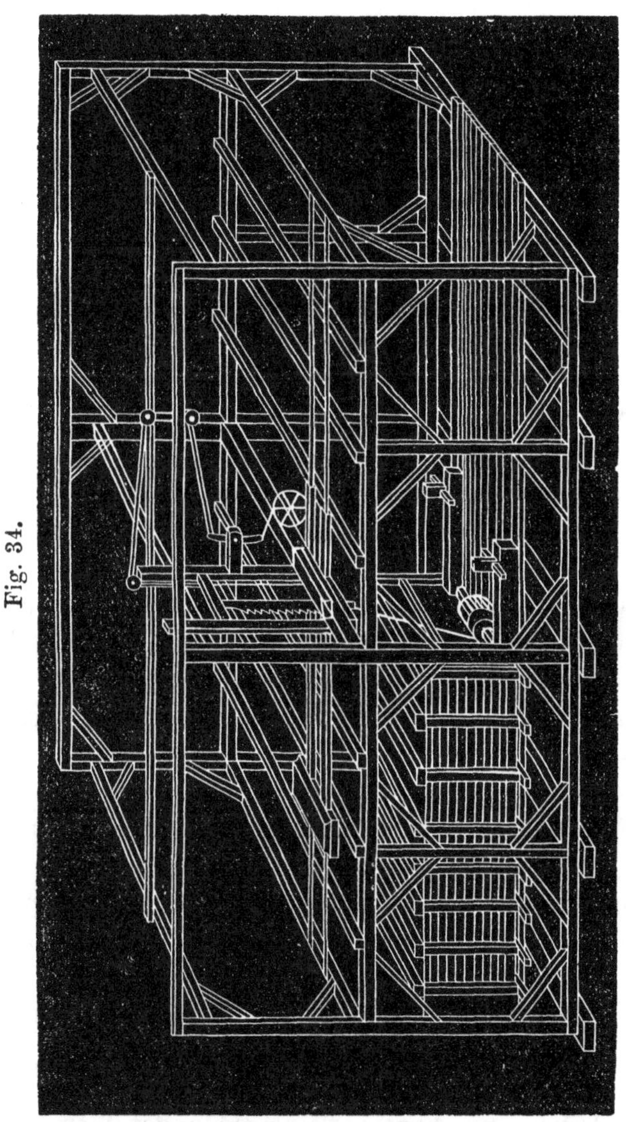

Fig. 34.

beam between them, as it would be in the way of the pitman. A pair of "streak sills" are framed on to the top of these posts, and three cross-timbers dovetailed into these streak sills, one at each end, and a very heavy stick, called the fender sill, across the middle. A pair of posts and a beam, of moderate size, are raised upon each end sill, and a large pair of posts, with beam to correspond, called the fender beam, raised over the large centre sill; this heavy centre bent is called the fender bent, because the fender posts, upon which the saw-gate works, are attached to it. A pair of plates placed upon the top of these second story posts will complete the frame ready for the rafters; girths are put in around the frame between the posts, both above and below, at proper distances, to which to nail the upright board covering; this is the simplest and cheapest frame that can be made for a common saw-mill. When the mill is intended for two gates, or more machinery of any kind, the frame must of course be larger, and other posts and beams must intervene between those mentioned, and another stick of timber must be framed in between the large centre post and the corner post, at a proper height for the log way, with a short post between it and the plate to receive the end of the extra beam. Some prefer placing studs between the posts, at proper distances, to nail to, and to put the boarding crossways and clapboard over these.

The flume may now be put in, and for a foundation for it and the machinery, place two large mud sills lengthways of the building running down stream past all the machinery, or as far as the building extends, and up stream at least several feet past the water-wheel, the further the better and under the bed timbers of the flume.

If the bottom be rock, these mud sills require only to be fitted to it, but if composed of loose stones or earth, flattened sticks will have to be placed under them, and sunk into the foundation, one across under the crank-shaft, and others at proper distances up and down stream; these, if the foundation be not safe, may extend across the whole width of the building. These mud sills must be placed at the proper situation and distance for the plumb blocks upon which the crank and journals of the water wheel revolve, and must correspond with the saw gate and carriage above. The bed timbers of the flume may now be placed with one end on these mud sills, about two feet back from the crank-shaft, leaving room for the baulk head posts, which must be framed down into the mud sills with strong double tenons, and either dovetailed and keyed, or well pinned, or both—as one end of the binders which hold the crank-shaft and wheel in place, is framed into these posts. The bed timbers run up stream to the head gates, or the whole length of the flume. Sills, posts, and caps are then framed of the requisite height and width to correspond with the baulk head posts, and placed upon these bed timbers, three or four feet apart, according to the strength required to withstand the head of water, and this portion will be ready for planking. The baulk head posts require two or three girths framed across between them, and a cap across their top: this cap must have a recess rabbited out of the under side and up stream corner, to form a bearing for the ends of the front planking, which go up and down instead of across, like the rest, and the cross girths for the same reason must be near the outside of the baulk head posts, to allow the ends of the side planks a hold upon these posts. The breast girths should be made with double tenons, one to go across the outside of the post, the other into a mortise about two inches from the face of the post, and fastened to it by a bolt; this secures the requisite space inside the posts, and also preserves the full strength of the girth.

This whole structure must be placed some distance up stream from the line of the fender sill and beam, because the fender posts are attached to that up stream side, and the friction guides, upon which the gate works, bring it still further up stream; and this point must be ascertained, and a plumb line hung from the centre of the lower bar of the gate, where the pitman is attached, will indicate the spot for the centre of the crank and shaft. The baulk heads must be placed some distance further off, so that the bearings cut in the plumb blocks and binders, for the crank and journal, may not be too close to the end. These plumb blocks and binders may be four to six feet long, and of a size proportioned to the bearings; a tenon may be made on the one end, and let into a mortise between the double tenons on the foot of the baulk head posts, and a facing may be cut in these posts, as deep as the dovetail facing of its own tenons, without impairing the strength, and the grip of the binder in this facing will help the tenon, and add to its strength and stability.

The baulk head must be stayed by long braces from the mud sills up into the posts, and pieces of good hardwood dovetailed into the mud-sills at the proper distance, with key ways mortised through the top, to key down the other ends of the plumb blocks and binders, which have a slit cut out of that end of each to straddle these pieces; another key is driven down between the shoulder of that slit and the upright piece to hold all solid against the baulk head.

This plan answers well enough where little power is employed, but where the mill runs strong, and strikes quick, it is not sufficient, and another pair of posts should be set up in place of the short starts, to hold the plumb blocks and binders. These should run up as high as the flume, and have a cap across the top to correspond with the flume caps; then a trimmer stick can be run in between these caps and the overlays supporting the floor and carriage, which answers the double purpose of holding the under works down, and supporting the upper. If the mill be large, these posts should be continued at proper distances upon the mud sills to the end of the building, and two trimmers placed along the top of these and each side of the flume, the whole length of the building. A good hardwood girth should be framed in between the baulk head posts and the next pair, about half way up, and right over the binders, so that an extra short post can be keyed in between it and the binder to hold all secure.

The long braces from the mud-sills will, in this arrangement, be put into these other posts, at or near the cross girth just mentioned, instead of into the baulk-head posts, as in the other case. The plumb blocks and binders may also be framed with a large tenon, to pass through a mortise in these posts, instead of the slit, &c., in the other arrangement, and a transverse key-way mortised through each, at these posts, to hold them firmly in place.

For planking a flume, it is necessary that the planks should be straight on the edges, although it is not imperative that they be jointed with a plane; a good saw joint, although it may not be perfectly water-tight at first, will soon become tight if properly made. There is more or less of foreign matter carried in solution along with all running water, even in the purest spring, the particles of which are so minute that they cannot be seen, but pass freely along with the current, where nothing obstructs it; but when arrested in their progress, they come in contact, adhere and accumulate, forming slime, moss, and incrustations upon the bottom and sides of the channel, which soon become water-tight. Now if a seam occur between the *square edges* of two planks, that seam will be as wide at the outside as the inside (the outside being dry it will be widest), and anything that enters the crack will pass through; but suppose you bevel the inside corner of these planks off so that the seam will be open on the inside an eighth of an inch, leaving a part at the outside square and close, then the seam will be contracting or funnel-shaped, admitting these particles to enter, but pinching them tighter as they pass through, and that seam will soon be, and remain, perfectly tight. We have seen a flume planked, only running off the inner corner of the seam with a drawshave, and spiking them on, the ends being sawn bevelling the same way, to leave the joints open inside. When the water was first let in it leaked like an old basket, but by shovelling in a lot of sawdust, it became quite tight in a few minutes.

The water ways to convey the water from the flume to the wheels, as well as the *gates* for opening and closing these, with the modes of working the latter, will depend upon the kind of wheel used, and more information may be found in our detailed description of the different wheels.

The water way for the old flutter wheel is a curve, the circle of which coincides with that of the wheel at the bottom, or termination, but gradually recedes from the wheel as it rises, to about the centre of the wheel, where this "compass fall" commences, and is the width of the "throat" away from the wheel. Thus only a portion of the water is intercepted by the first float or bucket, the rest passes on to the next one, which intercepts a little more, and so on until the last bucket catches the whole; this absorbs the force of the water gradually, and at the same time maintains its velocity and facilitates the discharge. The gate for this is a piece of thick plank the length of the chute, and somewhat wider, with the thick edge next the chute bevelled off and slightly rounded, to encourage the entrance of the water. A shank is fastened to this at the centre and well braced, a mortise is made in the end of the shank, and a lever passing through a roller, or having some other sufficient fulcrum, is placed with one end in the mortise of this shank, and the other end has the gate handle above attached to it; by moving this handle in one direction the gate is opened, by the reverse it is closed. This gate is generally made to slide upon the bottom of the flume, and shut against the lower breast girth, which for this purpose is made deep, and thinned away on the lower edge to the shape of the wheel, to allow it to approach close enough to the chute.

Nearly all the various iron wheels which are used in pairs and supplied by water through a curb or scroll between them, require a free communication with the flume, to maintain the full pressure of the water; and therefore the gate which controls these must be of large area, and cannot be made to slide upon the bottom or front of the flume, like the last. It is generally made nearly square, and hung upon an axle near the centre; one end being enough longer than the other to insure its inclination to shut. The long end is moved back into the flume, while the short one is forced into the spout, and the water passes on both sides of it; when turned back to its other position it closes the entrance; sometimes it is moved by a stem attached to the gate, at others by gearing attached to the axis. This large gate, although turned edgeways in the aperture, still obstructs the entrance of the water, and to obviate this it is sometimes hung by iron slings working in eye-bolts, and also by wooden slings attached to one or more rollers. In either case, these slings sustain the pressure of the water, and allow the gate to be swung off the entrance, leaving no obstruction to the water. All these varieties are worked by stems and levers, attaching them to the gate handle, which must be convenient to the saw and feed works, and varied according to circumstances.

The head gate is a necessary appendage to every flume, to admit or shut off the water, and especially so for a saw-mill, where so many contingencies are liable to occur, requiring the water to be shut off for their proper adjustment. This gate should be *large*, to admit a full and free supply, without losing any of the head, which will always be the case if the gate be too small. It should also *open easily*, because it often has to be shut and opened in a hurry, and with little help. To insure both of these requisites, it will be best to make two gates, joining at a post in the middle; if these be large, or the head of water considerable, a small scuttle may be cut in one or both of these, and covered by a small gate, which is easily opened, to allow the flume to fill up (the gates below being closed) when the main gates can easily be lifted. They are generally made to slide up and down the baulk-head, which sustains them, and moved by a lever, a rack and pinion, or windlass, attached to the upright stem; this stem, however, is liable to be frozen solid into the ice, and difficult to loosen in an emergency, and for this reason some prefer to hang the gates upon iron hinges, at the upper edge; they are then totally under water, and out of the way of the ice, except a chain which is attached by a ring or eyebolt to the lower edge, to open or shut it. There is scarcely any part of the whole structure so liable to be neglected and out of order as the head gate; and no part so likely to be wanted on a sudden emergency. A gate may get wrong and refuse to shut, or a stick or other obstruction may get into it, or a break may occur somewhere, or a person fall in, or other contingency requiring the water to be shut off instantly; therefore it should be made and kept in a condition to be easily and quickly shut down; and again, it is frequently necessary, when the water is thus shut out, for a person to go down into the flume, sometimes under the ice, or into the wheel curb or spout, where a speedy retreat is impossible, and certain destruction the consequence, if the head gates should give way.

Three instances of this kind have occurred near where this is written, one the season before last (1867), on the Chateaugay River just above the celebrated falls. A man named Josiah Bull crept into the curb supplying a pair of the Hildreth wheels, to fix something inside; the head gates, which were known to be insecure, gave way, and although he was very active, and said to be the strongest man in town, yet so great was the velocity and force of the rushing water, that he was unable to escape. The owner of the mill had been in the same curb a short time previously, and remarked to Mr. Bull when he came up, on the danger of the situation, if the head gates should have given way. Mr. Bull answered, "You would certainly have got a ducking; but I would have cut the planking of the curb outside and let you out before you drowned." The owner's presence of mind so far forsook him, when the dreaded catastrophe actually occurred, that instead of doing this he ran to the top of the bank, which was very high, calling to Mr. Bull's brother for help. When the brother got there he cut the curb open directly, but it was too late, and Josiah was dead when taken out.

Another instance happened, when five men and boys lost their lives. This flume was decked or planked overhead, the head of water being higher than the lower story of the mill, and a ventilator about sixteen inches square was carried up to a height a little exceeding that of the water, to relieve the strain or concussion of the water when the gates were closed. The water was shut off at the head gates, and several men were at work in the flume, some other men and boys had also gone down into it from curiosity, when the head gates gave way, and instantly packed the whole flume full of water, driving several persons up the ventilator who happened to be near the bottom of that tube, and were thus rescued, while five of the number perished before any assistance could be rendered.

The other incident referred to had a less tragic termination. We were gearing a new double saw-mill, which was built at the down-stream end of an old one; the flume in the old one had lately been renewed, and we continued it further on to supply the new one, leaving the old head gates in the mean time just as they were. We had started one saw gate in the new mill, but the swing gate closing the entrance to the wheel curb shut so suddenly and so tightly, that the vacuum collapsed, or sprung a piece of plank loose inside the throat, and the owner shut down the old head gates, which had not been renewed when the flume was, and persuaded an active young man, a cousin of mine, to enter and fasten the fugitive plank. He had just slid himself in feet foremost, and flat on his back, when the head gates gave way with a tremendous crash, and Mr. B—— scrambled up the ladder and left the young man to his fate. His own activity, however, and the length of the flume, enabled him to get out and clutch one arm around the post, to which the gate was slung, just as the water struck the baulk head; and being in a corner, the whirl or eddy of the water struck his legs and body, and carried him round and round this post as it filled the flume, bearing him to the surface, when my brother, who had run to the rescue, caught and dragged him out, his clothing and skin pounded and peeled by the square corners of the post and contact with the planking. However, he escaped with his life, and is now foreman of the Lake Champlain Steamboat Company's Works at Shelburn Harbor, Burlington, Vermont

But this is another digression, from which, however, some useful hints and suggestions may be gleaned and appropriated by some persons, which I will plead as a partial excuse, and endeavor to get back to the machinery.

If the crank shaft be iron, the wheels, whether one or a pair, should be bored out to fit the shaft, and fastened in their places by a steel key; but if a wooden shaft, it must be made large enough to fill the opening through the wheel. If the original stick be not large enough, it can be pieced up to the size, and turned off to the exact shape of the opening; the wheels must be wedged on solid and true, to run tight to the curb or chute without leakage or friction. When the flutter wheel is used, three rims are generally made around the shaft, with openings into which the floats are fitted, then an iron band of the exact size is heated to enlarge it and make it slip on easily, and shrunk on around each rim, to strengthen it and hold the floats in their place. These rims are made of hardwood plank, as many pieces as there are floats in the wheel, and each piece being an equal portion of the whole circle, with a tenon on the narrow inner end to go into a mortise in the shaft, and a piece cut out of one corner of the wide outer end to admit the float. The floats are about four inches wide by two and a half thick, the face next the water flat, and the back rounded off to a half circle; this form helps to shed the back water. These wheels are sometimes made altogether of iron, with the three rims of cast iron, keyed to an iron shaft, and the floats of bar iron, with a half round piece of wood fastened along the back of each, to improve their shape.

Formerly a simple crank of wrought or cast iron was used without any balance, the wrought crank having a square or flat shank let into the wooden shaft; this is done by cutting two slits out of the wood, the length and width of the shank, each slit or recess at *right angles* with the other (not opposite). The shank is laid in through one of these to the centre of the shaft, and three iron bands put on. Then make three hardwood keys for each, of such breadth and thickness as will exactly fill each recess; place two keys in each, with the thick ends behind, and enter the point of the third one between these two, and drive it home; the centre key should be left longer than required, as the battered end requires frequently to be cut off and the driving resumed. A ram made by hanging a long hardwood stick by two chains, and in the right position to strike these keys, when it is drawn back and pushed forward, will be the best driver, as it should be very tight. The cast-iron crank has wings like a gudgeon, and is laid and secured in the shaft in the same way.

Some kind of balance is now almost invariably used, with reciprocating saw movements of every description. To understand the benefit of the balance upon a crank, we must consider the great weight of the gate, the pitman, and feed works, with all the irons connecting these, together with the saws and their hangings, and that the combined weight of all these is attached to the crank. Now, when the crank is set in motion, all this weight has to be alternately raised up and jerked down again, and that with a speed equal to the motion of the saw, and this velocity is so great, that when the gate is going up, it would continue right on up through the roof if not held back; but this upward stroke must not only be instantly stopped, but reversed, and the whole fabric sent *down* again with the same velocity and force. Here again, it must be stopped and reversed, and so continually, for every stroke of the saw. This uses up a large proportion of the power applied to every kind of saw gate, and shows the expediency of making the gate, pitman, and every other part of these reciprocating movements as light as possible, to be safe; and then balance the weight of these, as near as may be, by a balance weight attached opposite the crank pin.

It is by dispensing with the weight of the gate, or sash, as it is sometimes called, and the corresponding diminution of weight in the pitman and other fixings, which the arrangement admits of, that enables the "mulley saw," *i. e.* saw *without* a gate, to strike so much faster, and do so much more work with the same power, than the English Gate.

The mulley arrangement, however, is subject to the very serious objection, that no provision is made for straining or tightening the saw; the saw has therefore to be stiffer and of course thicker, on that account, than the common English gate saw. This extra thickness cuts a wider saw draft through the log, and thus causes both a waste of timber and a waste of power to make this greater quantity of sawdust. The mulley must also be hung and maintained accurately in every respect, and set and filed perfectly true, requiring a first class sawyer to keep it in proper working order. The ordinary catch sawyers, that will keep and run a saw well enough when strained tight as a fiddle string in a heavy gate, would make bungling work with one of these.

It will be found true therefore in theory as well as in practice, that although it is not best to hang a saw in a gate that would require two men to lift it, and with a pitman as heavy as the gate, neither is it best to hang it without a gate altogether, and with only a strip of two inch plank for a pitman. Common sense would seem to indicate a medium somewhere between these two extremes; and instead of straining a saw with the whole strength of the stirrup irons, or running it without any strain, we have always obtained better results by using a light gate, made of light, elastic wood, and sufficient to give the saw a reasonable degree of tension.

Another point upon which a great diversity of opinion exists is, the proper length of the crank. A certain bias obtains among millwrights and mill owners in favor of *short* cranks, for the reason that a quicker stroke is given by these, than by longer ones, other things being equal; it being a popular boast to have a "smart" or quick striking mill.

This prejudice does not exist to the same extent among sawyers; experience soon teaches them that the faster any gate strikes, the more trouble it is to keep things in order; and, also, that the quantity sawed does not depend upon the number of strokes, but upon *the velocity of the saw.* In other words, two strokes of a twelve inch crank are equal to three of an eight inch one; two strokes of a fifteen inch crank are equal to three of a ten inch, performed in the same time. If this were *all* they would be equal, but the three strokes have six stops and reverses, or six corners to turn, while the other has only four; and it is these reverses, not the cutting of the wood, that use up the greater part of the power applied. To explain this, we will suppose an engine upon a railroad, with a light train, or rather attached to a single car, and required to move that car over so many thousand feet of track in an hour. When the engine started it would jerk pretty hard upon the coupling to start the car, after which a slight steady pull upon the coupling would soon carry it over the given distance. But suppose, that instead of being allowed to continue right ahead along the line to complete the given distance without a stop, it should only go ten feet and then reverse, and go back over the same ten feet, and so continuously until the given distance was made. We can scarcely imagine the increase of power which this would require, or the excessive wear and tear it would involve upon the whole concern, particularly the couplings and buffers, and how much it would increase all the disadvantages to shorten the range of movement still more, or how much it would relieve it and assist in accomplishing the given distance, to lengthen the range, say to fifteen feet, which (like the lengthened crank) would make two reverses equal to three.

But this is a mode of reasoning for the long crank, which all sawyers may not be able to appreciate; I will therefore give others in its favor for their (the sawyer's) especial benefit. When a saw has a long range of stroke, it moves faster in proportion as the range is lengthened, and every tooth is made to cut nearly its equal proportion along the whole face of the saw; this wears upon all the teeth nearly alike, and the saw is easily kept straight and in order, besides which every tooth has a chance to discharge its sawdust either above or below the log, at each stroke. But with a short stroke the middle teeth must do nearly all the work, the end teeth never entering a small log at all, while the middle teeth are never out of a large one, and have no chance to discharge the sawdust, which clogs and heats the saw. The middle of the saw in this case, does all the work, and has a constant tendency to wear hollow in the middle, while the end teeth, which get no wear, require constant stripping, gumming, and filing, to keep them in line with the middle ones; and where the short crank is used, and more especially when the logs are of small size, sawyers are in the habit of cutting off several of the end teeth, to avoid this extra labor.

The pitman for an ordinary English gate is generally about six by nine inches at the lower end; and four by seven at the top; a piece about two feet long is left full size at the butt or lower end, with a hole through this part about nine inches from the lower end, the size of the crank pin or wrist. This hole is extended up in the form of a mortise, about eight inches long, and square at the upper end, and a box fitted into it circled out at the lower end to fit the crank-pin, and with a space left at the upper end for a key to be driven through to tighten the box as it wears loose. The head of this key may be left long to drive again, but the point must be cut off close every time, as it would interfere with the balance when it comes round.

An iron strap or clasp is put around the pitman at

the centre of the box to keep it in place, and strengthen the pitman. This clasp is made in two pieces, to admit of its easy removal, the screwed ends of one piece passing through the other, and tightened by nuts on the outside. Another *whole* band is sometimes heated and shrunk on the extreme lower end, but this leaves the end too square to work well, if there is any backwater; and it is better to cut the end down to a kind of thin point, and put a bolt through near the point, and two others up near the wrist hole; it is sometimes necessary to bolt through above the key mortise also.

The upper end is finished by commencing at the full width of six or eight inches, and about twenty inches from the end, and tapering it down from this point on both sides to the thickness of the noddle pin at the end (working from a centre line struck upon both sides); it is left the full thickness the other way, and one or more bolts are put through to prevent its splitting. The noddle pin is fastened on by an iron strap, about four inches wide, bent over it and running down along the tapering edges of the wood to the wide part. This strap has a mortise through each end about an inch wide and three or four long, and a mortise is made through the wood to correspond with these; three iron keys are made to fill these mortises, two of them are placed in the mortise and the third one driven between them, the mortise in the wood being lower than those in the iron, to cause it to draw and tighten. A screw bolt is put through near the top to keep the strap in place; the hole through the wood being enlarged downward to allow the strap to be tightened by the keys. The wood should be thinned away from the end of this strap downwards, to lighten it; it may also be thinned away from above the crank key upwards, but it should be the full width at about one-third the length upward, as the swing of the crank might cause it to spring. As the strain upon it the other way is but trifling, it may be bevelled from the centre lines almost to an edge.

The pitman is generally made of hard heavy wood, but a thrifty little spruce, grubbed up by the roots, makes a very good one; it being light and stiff, and with the root for the lower end, and well made and ironed, will answer a good purpose. The noddle pin being secured by the middle to the pitman, as described, a pair of wooden boxes, about six inches long, are fitted on to each end of it, and these boxes are fitted on to the under edge of the lower cross-bar of the gate. An iron strap passes over this gate bar, and down through the ends of each pair of these boxes, and the nuts on the lower ends of these straps being tightened, secures the whole to the gate, while the round ends of the noddle pin, working in these boxes, allows the pitman the necessary play. The length of the pitman must be graduated exactly to agree with the carriage, fender posts, and everything else affected by it above. The simplest and safest way to get the length, is to raise the gate up to the highest point of its intended stroke; and turn the crank up also to its highest range, and take the measure with a pole. Of course, the lowest range of both being found, if more convenient, will give the same result.

The above is a common way of making and attaching the pitman to an ordinary crank, driven direct from the water-wheel without gearing. For a geared mill, driven by a belt, or otherwise, and clear of the water, the lower end must be provided with metallic or wooden boxes, easily removed and replaced, and with some provision made for oiling; almost every millwright of note has a different way of attaching these.

The gate or sash consists of two upright stiles, and two cross-bars, between which the saws are hung and strained. It is generally made of hard wood, although some light wood that is stiff and elastic, such as spruce, is preferable, particularly for the upright stiles. There are many different ways of putting these together, varying with the notions of the maker, and some of these are very fanciful; some compose the cross-bars of several different pieces framed together with braces and slings and other contrivances to combine the requisite strength with lightness; these are generally better examples of ingenuity and patience, than of utility. The saw is attached to the lower bar by two iron stirrups around the bar at the centre, and between the two straps of the noddle pin boxes. These stirrups should be long enough to come about three inches above the bar, and should be slipped on it before the gate is put together. The lower end of the saw is slipped in between these, and held by a "lug-pin" put through a hole drilled near the end of the saw for its reception. The upper end is secured in the same way by two stirrups passing over the upper bar; these, however, are open above, the ends being screwed and passing up through two strong pieces of iron, with nuts above for tightening the saw.

A strong spring, three feet long, or more, is fastened by one end to the top bar of the gate, the upper end being attached by a mortise and pin to the point of the feed pole. The gate is attached to the fender posts by four pairs of wooden boxes, one at each corner, clasped around the friction rods, or guides, and bolted to the gate. These guides are of iron, of sufficient length for the intended stroke, turned or planed true, some round, others square, and bolted firmly to the fender posts.

The fender posts are of large square timber, and are

let into a notch cut out of the fender beam and sill about two inches deep, and wide enough to admit the post, with a dovetail key on each side for its ready adjustment and security. The posts have a notch cut out behind, opposite the beam and sill, four inches deep, and two inches wider than the depth of the beam and sill. A hook tenon is made at the top of each of these notches, by leaving the first two inches cut out of the post two inches lower, and cutting out the next two inches clear to the highest mark. A mortise is made in the top of the fender beam and sill, at each place, two inches back from the face, and two inches wide and deep, to receive these hook tenons. The fender posts are then raised up until the tenons will slip back over the top of the beam and sill to their respective mortises, and dropped down into their places; a key is then driven in upon the shoulder under the beam and sill, which keeps them firmly in position.

The feed pole, before mentioned, is ten or twelve feet long, about three by four inches at the big end, which is put through the rocker, by a tenon, or secured to it by joint-bolts; the small end about two by three inches, is put through the feed spring on top of the gate, by a loose tenon, with a pin through to allow it to play. As this small end is carried up and down by the motion of the gate, it imparts a reciprocating rocking motion to the rocker. The rocker is a piece of wood about eight inches square, and of any convenient length, with journals and bearings at each end to admit of motion and keep it in place. It is placed overhead at a convenient distance beyond the wrag-wheel, and besides the feed pole already mentioned, it has four standards mortised into its under side, each being two inches square, and two inches apart. These are connected together at the other (lower) end, for greater strength, and one end of the connecting bar is slipped through the two inch space between these standards, and a short two inch piece or pin thrust through the end of that bar, at right angles, and in the contrary two inch space between the standards. This allows the end of the connecting bar to be let down to the end of these standards, which gives it a very long stroke, or taken up close to the rocker, which gives a very short one; and by fixing a lever overhead, with the short end connected with this movable end of the connecting bar and the long end at the fender post, the feed can be varied by the sawyer at will.

The evener for an English gate is generally a lever of the first kind: that is, the fulcrum or bearing is passed through it near the middle, while the hand that pushes the wrag-wheel around is attached to the lower end. For a gang, it is generally a lever of the second kind, the fulcrum being at the lower end, and the hand attached near the middle; the power, in both cases, being applied to the upper end, through the connecting or feed bar.

The wrag-wheel is a wooden wheel about four feet in diameter, with iron ratchet teeth or notches around its circumference for the hands to catch against and push it around. It is attached by four or six arms to the wrag-shaft, which lies across under the carriage. If this shaft is of iron, it has two pinions keyed to it, corresponding with, and gearing into, two cast-iron cogged racks placed along the bottom of each side piece of the carriage. These racks are placed so as to fit close along the *inside* of the ribbons or ways upon which the carriage runs, and thus serve the double purpose of feeding the carriage forward, and gigging it back by the pinions, and also of keeping it true and parallel in its motions.

If the wrag shaft be of wood, it may have cogs mortised into it in two rows, two inches apart, to straddle the ribbons, six or eight in each row, and so stepped off that the cogs in one row will be exactly in the spaces of the other row. Thus the two separate rows of six or eight each, are equal to one continuous row of twelve or sixteen, and being divided in this way leave sufficient space and whole wood between. These cogs must not be left nearly square at the ends like ordinary cogs in wheel work, but having to step into mortise holes in the straight line of the carriage, they must be of a peculiar curve, and terminate almost in a point. To mark this shape, step the rough cogs upon the pitch line to divide them equally, and mark the outsides at this line; then wind a line around the shaft, and fasten a scratch-awl in it so as just to reach the mark on the further side of the cog; begin at the root of the cog and mark as the line unwinds to the point; move round to the next cog and mark the same, until one side has been marked; then reverse the string and mark the other side the same way. A piece of small wire is handier and safer than a line, as it will not stretch, and by sticking a tack through a loop in the other end, it is more easily managed.

With exactly the same step as these cogs, the mortise holes for them to catch into the carriage must be stepped and laid out. And as there is nothing to keep the carriage true and parallel, the two inch space between the two rows of cog-holes must be rabbited out to the depth of three-eighths or half an inch, to correspond with the ribbons on which it slides.

A better, because a more accurate and durable way of making this wooden shaft and gearing, is to put the cogs into the carriage, a row on each side of the ribbon, and cut recesses out of the shaft for the cogs to step into, leaving round ledges between these recesses for the cogs to catch and work upon. The *shaft*, in this case, must be enough larger to correspond with the step of the *cogs* in the other plan; or in other words, the working surface of the rounds in the shaft must be the same distance apart as the working surface of the cogs in the carriage. These rounds are sometimes turned in a lathe, and fitted around the shaft, with an iron band around each end to keep them in place, like an old-fashioned trindle head. In this way they are easily removed, which is a desideratum, as one set of cogs in the carriage will wear out several sets of rounds.

The carriage consists of the two side sticks already alluded to, about five inches deep, and eight wide, and about twenty-four feet long. This is sufficient for any ordinary length of log, up to eighteen feet, for when timber longer than this is to be sawn, it is more convenient to extend the length by attaching a movable piece which can be laid aside when the common lengths are resumed. These side sticks are connected by a cross-bar mortised through each end; the tail block is also pinned firmly down upon the end of these side pieces, but the head block is only notched down an inch or two, leaving a shoulder inside at both ends, to keep it from shifting endways; this leaves it free to move the other way, to suit the various lengths of logs. These guiding shoulders soon get bruised and worn, both on the carriage and head block, and allow it to vary, and sawdust and bark, or slivers get between the block and carriage, and impair the accuracy of the work. To obviate this disadvantage, the sides of the carriage are sometimes made with a high ridge along the centre, and bevelled off straight and equally on both sides. To do this, strike a line along the top at the centre, and another line along each side, say two inches from the top, and dress off from the centre line to each of the others; then notch out the ends of the head block to fit this saddle, and its own weight, with the weight of the log will always keep it in its proper place, while the incline of the surface will keep it free from sawdust or other accumulations which would disturb its accuracy.

The track upon which the carriage moves is made in different ways; a simple way is to take pieces of hard wood about three inches thick and five wide, and key them into the sills and overlays on their edges. This is done by striking a line the whole length of the mill, to mark the *outside* of each line of ribbons; another line is struck *inside* of each of these at a distance of about five inches from the others. The first (outside) line is sawed down plumb, and straight with the line, two inches deep, the other (inside) is sawed a little under and dove-tailing, to admit a key to the same depth. The piece is cut out with an adze, and the rough sawed pieces for the ribbons placed along the outside of these gains, and keys driven in along the inside in every gain, to tighten the pieces. Two lines are then struck along the top of these, two inches apart, to mark the ribbons, and the surplus wood dressed off. The line is then stretched along the side, at the height of the ribbon, and raised crowning a little at the middle, and when jointed off to this last line the ribbons are completed.

Another plan is to key strong foundation pieces into the overlays, and dress them off true on top, then gauge the ribbons and dress them off to the proper width and thickness, and pin them on to these pieces by lines struck the whole length. When these ribbons get worn and out of true, they can easily be stripped off and re-newed. The new track should always be made a little high at the middle (the saw), as it works better, besides it has a constant tendency to work down at this point.

The *evener*, when hung by the middle, has its fulcrum supported by a short piece of plank on each side; these are either mortised into the fender post, or into a separate post put up for that purpose, as may best agree with the line of the wrag-wheel.

Below the movable hand already mentioned, is another stationary one, attached by the back end to the post, by two staples or eyes, the other end or point resting on the wrag-iron. Its purpose is to hold the wrag-wheel from turning back by the recoil of the wheel and the force of the saw, while the feeding hand is drawn back to make the next stroke. These two hands should be connected together by a small chain, or otherwise, and both attached to the end of a light beam above, which is hung by the middle, and the other end should extend over to the outside of the carriage. To this end is hung a piece of scantling heavy enough to overbalance the weight of the hands and chain; this scantling stands upon the floor, beside a hole, when the hands are on or working, leaving a little slack to the chain. When the carriage is fed forward to the proper extent, a piece attached to it for that purpose, pushes the scantling into the hole, and thus lifts both hands clear of the wrag-wheel, and leaves it free to be gigged back. After it is gigged back until the saw is clear of the log, the drop is raised up to the floor, and this movement lets the hands down on the wrag-wheel again.

The tail block is generally made of a single piece of hard tough wood, seven or eight inches thick, and eighteen or twenty wide; a piece is cut out along the upper corner, about two inches deep and six wide, upon which the end of the log rests; a single dog, with two tines to drive into the log, is hinged on behind the ledge by two eye-bolts passing down through the block, and secured below by two nuts. The head block is similar to the other, but a little wider, and a notch is cut out in the middle to allow the saw to come back clear of the log; it has two dogs, one on each side of the recess for the saw. This recess cuts the block so nearly through, that it is liable to be split by turning down a large log upon it; to strengthen it some builders dovetail pieces across it, others put several large pins through; perhaps the best way is to put iron bolts through, as these can be tightened as the block shrinks by seasoning. The other plans do not admit of tightening, and rather encourage it to check. The space between the dogs and the recess in the block is generally made wide enough to admit of a span of saws being used, as these will cut about double in the same time that a single saw will cut, while one-half of the time occupied in gigging back and setting is saved. And the span is generally used when the kind of stuff made admits of it.

The process of gigging back is next to be considered. There are many ways of accomplishing this, varying according to circumstances. If there be an edging circular saw in the mill, or anything else requiring a horizontal motion, both the gig motion and that for hauling in logs may be got from the same horizontal shaft, by a belt or chain, to be tightened or slacked by a movable tightener. If no such motion be required for other purposes, and a gig wheel has to be made for that express purpose, then it is best to put a small water wheel on an upright shaft, under the wrag-wheel; the shaft to reach up and come in contact with the side of the wrag-wheel. Both shaft and wheel must have a particular bevel, proportioned to the relative diameter of each, in order that the whole surface of contact in both may move at the same velocity. If the bevel be too much or too little, the upper and lower edges of contact will move with different velocities, and cause a constant slipping and grinding of the surface, which increases as the pressure is increased, and soon wears the face of both wheel and shaft so uneven as to be unfit for use.

A failure to proportion the bevel of this wheel and shaft to each other, is the most common defect met with in ordinary cheap saw-mills, and has contributed more than anything else, to prevent this mode of gigging back by friction from coming into more general use. It certainly is the handiest and cheapest of any yet in use, and if the wheel be bevelled to half the diameter of the shaft, and the shaft bevelled by half the diameter of the wheel, it will be more durable.

To do this, place a straight edge or line across the face of the wheel at the centre, then measure out from that line one-half the diameter of the shaft, and at that point place a sweep, and bevel the face of the wheel until the sweep agrees with the face all around. If the wheel be on the shaft, and the shaft in the way, measure *its* diameter, and allow for it, and make a mark around the shaft with that allowance to set the sweep. With the shaft, proceed the same way, measure out from its centre, and in a line with its shaft, one-half the diameter of the wheel, and at that point set the sweep, dress off the same as with the wheel, until the sweep agrees with it all around, and the bevel will agree with the wheel. Both surfaces should be turned, to insure their accuracy.

This bevelled friction wheel and shaft work upon the same principle as the bevel wheel and pinion in cog gearing, and are governed by the same laws. Any kind of water wheel may be put upon the lower end of the friction shaft, to give it motion; the wooden centre-vent or scroll wheel being as convenient and cheap as any. This is made by squaring the end of the shaft to six sides, and bolting, or otherwise attaching a short piece of plank or float to each square, with the outer ends pointing in the direction the wheel is intended to turn. The scroll may be made of four pieces of timber of the proper depth, framed together, with one side falling out

Fig. 35.

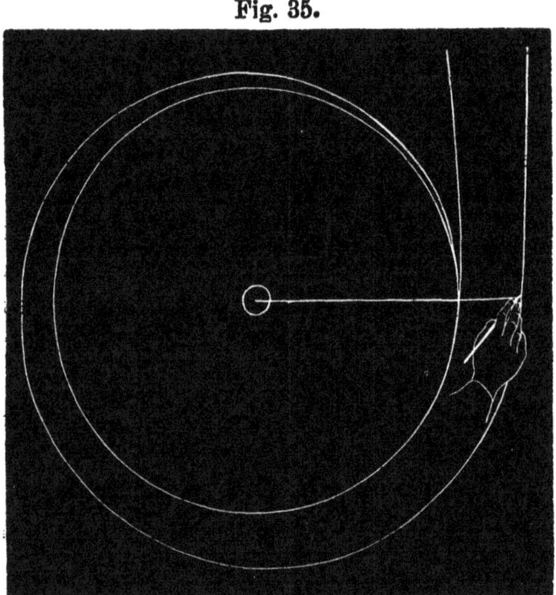

from the square, and leaving an opening for the admission of the water; the scroll mark is struck around upon this, cutting a portion of the circle out of each piece, and the corners are filled out to the same line. A bottom and cover are spiked on, each having an opening around the shaft to discharge the water.

The following method of making a scroll is in some

respects, preferable: make a square platform of plank for the bottom, a little larger than the outside of the scroll; make a centre mark in this at the point which the centre of the shaft will occupy; then with a radius or sweep from this centre draw a circle to correspond with the outside circumference of the wheel (floats), and from the same centre draw another circle about two-thirds of the diameter of the wheel, for the opening for the shaft, and discharge water to pass through. Now mark off the width of the throat for the admission of water, commencing at the circle for the wheel, one side of this throat is the commencement of the scroll, the other side its termination. (See Fig. 35.)

To mark out the scroll, make a small wheel, the circumference of which is exactly equal to the width of the throat; the wheel may be tested by measuring round it with a string, the length of which is equal to the width of the throat, or by making a mark on one side of the wheel and placing that mark at one side of the throat, and rolling it (the wheel) across the space to the other side. If the mark, when the wheel has made its revolution, comes exactly on the other side mark of the throat, it is right; if it overruns that, the wheel is too large, if it falls short, it is too small. Fasten this wheel upon the centre of the bottom, and attach a line or a small wire to the edge of it, stretch the wire tight and fix a scratch point in it at the *outside* line of the throat, now draw the scratch round, keeping the wire tight, to the place of beginning, and the point will mark the scroll, terminating at the inside. Or, you may wind the wire round the wheel, and begin at the *inside* to mark, allowing the wire to unwind as you describe the scroll, which in this case will terminate at the outside; for this, like other good rules, will work both ways.

Now lengthen the wire an inch and a half, or two inches, and describe another mark parallel with the first, and cut the space between, out to the depth of half or three-fourths of an inch. Cut planks of a length equal to the intended depth of the scroll, and circle them like staves, then fit them into this space all around until the scroll is complete, and lastly, make a top or cover the exact counterpart of the bottom, and place on to these staves, and bolt or otherwise fasten the top and bottom together at each corner, and the curb or scroll will be complete. The throat, or opening for the admission of water, is connected by a spout with an opening in the flume, which is covered by a gate inside. The lever, or gate handle, for moving this gate should be brought conveniently to the wrag-wheel, and another lever with a short purchase, to throw the head of the gig-shaft *in* and *out* of contact with the wrag-wheel, should be within the sawyer's reach at the same place; as it is most convenient when he can have one in each hand at the same time.

The pull wheel, or bull wheel, as it is generally called, is similar to this gigging apparatus just described, the principal difference being that where the gig wheel has cog-gearing upon its horizontal shaft, to run the carriage back, the bull wheel has a chain winding around that shaft to haul the logs in. There is also this further difference, that while the gig wheel moves its load upon a smooth and level track, the bull wheel must haul its load upon a rough track, and up a steep incline. The wheel upon the horizontal shaft should be larger in diameter for this reason, to increase the purchase, and if driven directly from the upright shaft of the water wheel, both wheel and shaft should be furnished with cogs to match; because the motion obtained by friction which answers admirably for a light free movement, is not reliable where a power like this is required.

The slide upon which the logs are drawn up out of the water, is sometimes made of two large sticks flattened upon two sides; these are placed on edge, at a distance apart equal to the intended width of the slide, and bars or slats mortised through them at convenient distances, upon which a flooring of plank is pinned or spiked. The butt or strongst end is placed upon the sill of the mill, the small end running down obliquely into the water, and either resting on the bottom, or upon an a abutment of stones or wood prepared to receive it. This slide is frequently made altogether of round sticks of timber, a large stick being placed at each outside, and the space filled with smaller sizes, which makes it high at the edges to prevent the logs falling over. These sticks are all mortised to correspond, and slats driven through them, the outside ones pinned through the slats, as in the other case, to keep all secure. This makes a stronger slide than the other, but *either* is strong enough to carry any load of logs, while *neither* is sufficient to withstand the ice in many situations, if the water is drawn down. Where such an occurrence is anticipated, the slide must be well supported by posts, or otherwise, as without this precaution, the slide might be broken or the mill itself disturbed by the oblique pressure of the ice.

All the various parts and pieces comprising the machinery of a saw-mill should be of good material, and accurately made and combined, no need of fancy finishing anywhere, but the movements adjusted correctly by the plumb and level. The water-wheel (crank shaft) level, with "dead wedges" (wedges the same thickness at both ends) between the binders and the plumb-blocks, so that the binders may be keyed down firmly without pinching the shaft, the saw-gate parallel with that shaft, and plumb above it; the centre of the gate at the noddle-pin and stirrup exactly over the centre of the crank-wrist. The gate must be hung plumb, each way, upon the friction rods or guides; to insure this, these guides must be placed upon the fender posts by a line, both ways, to get the two upon each post perfectly true with each other, then the boxes that attach the gate to these guides should also be placed by a line to get them parallel and true with each other. When this is done, fasten one fender post solidly in its place, the rods plumb each way, but do not key the other fender post until the gate is set up, and the boxes are put on to attach it to the guides; tighten up the screws of the boxes, then key the other fender post, now slack all the boxes and place dead wedges between to give the necessary freedom to the rods. After tightening all up again, the gate will be ready to run. The noddle-pin boxes, which attach the gate to the pitman, must also have dead wedges placed between them in order that the connecting straps may be firmly screwed up, and at the same time admit of the necessary freedom of motion.

When all these various points of motion are right and tight, the gate will play up and down freely, and without any jolt or noise, in reversing at the upper and lower extremity of the stroke; but if it binds and is hard to start, then it is too tight somewhere, and the tight spot must be ascertained and slackened, and thicker dead wedges introduced and all made fast again. If it jolts and makes a thumping noise at every revolution, then it is slack somewhere, and to ascertain the point, take a short lever, pry open the gate (sawyers generally catch a tooth of the saw with the mill bar on top of the log for this purpose), and move it up and down as far as the slack play admits, another person examining the different joints will easily detect the loose one, which must be slackened, the dead wedges taken out and dressed thinner, then replaced and tightened as before.

When the lower end of the pitman wears loose upon the crank wrist, it is tightened by simply driving the key above the box. If this key is mortised through crosswise of the box, the projecting point will do no harm; but if it passes directly through on the top of the box and in the same mortise, then the point must be cut close off every time it is driven, otherwise the projecting end will come in contact with the balance as it comes around, and in this way the pitman sometimes gets broken.

To hang the saw, place it in position in the stirrups, with the upper and lower ends at the same distance from the friction rods or guides upon which the gate works, then feed the carriage forward its full extent until the tail block comes in contact with the saw, and mark this point permanently upon the tail block, or better on the dog, for after use in hanging the saw. Now gig the carriage back to the length of the saw logs and fasten a line upon the mark just made upon the dog, stretch this line along both sides of the saw, both at top and bottom, and shift and secure the ends of the saw by wedges or otherwise in the stirrups to range exactly with the line. Some prefer a short straight edge to the line as being equally true and more expeditious. This is applied first to one side of the saw then to the other, and the wedges (or screws) manipulated until the straight edge points equally from both sides and both ends of the saw to the mark on the tail block and dog. Now see that the saw has the requisite "rake," that is, that the lower end be hung back three-eighths or three-fourths of an inch further than the upper end, according to the quantity of feed carried; this rake is required to enable the log to feed forward against the saw, without the teeth catching or cutting in the upward stroke, and should be nicely graduated to the feed. If a saw has too little rake, for the feed put on, then the teeth will catch during the upward stroke, and as they cannot cut on account of their position and hooking shape, they catch and lift upon the log, jerking it loose and disturbing its position, besides greatly retarding the velocity of the motion. If too much rake be given, the saw will recede too far from the log in its upward motion, and be partly down again before the teeth touch, the under ones never touching at all; and the upper ones having to do all the cutting, will strike in so sudden and deep that the whole cut will be rough and coarse, and the lower edges torn and splintered, making a rough and ungainly description of lumber even out of the best of timber. Too much rake also produces a concussion or shock at each stroke of the saw, which is very disagreeble, besides being injurious to the saw and all the movements and their fastenings.

This subject, of the proper rake for saws hung in a gate, either singly or in gangs, is very important, and is much neglected and but indifferently understood by many otherwise good sawyers. It is quite common to meet with sawyers who file and set a saw well, and hang it to range right in its track, and yet have so limited an idea of the importance of this subject, that they use but one rake for all kinds of wood soft or hard, and all kinds of saws long or short, and every length and strength of stroke. When a saw is hung to range plumb and true, both ends maintaining the same parallel cut, and with the proper rake so that every tooth cuts its equal share, then it will work with less set, having no sideway vibration, less waste of wood and power, the saw draft being narrower, and less filing, the wear and tear being equally distributed among all the teeth. At the same time the lumber cut will be much truer and smoother, requiring but little labor to plane or dress it. Surely these considerations are sufficient to induce sawyers and particularly owners, to pay more attention to this subject.

Filing and setting the saw is also of great importance. The saw cutting only during its descent wears the points of the teeth mostly on the under side; this, particularly in knotty and hard wood, gives a constant tendency to the teeth to turn up at the point; whereas, to cut easy, the sharp peak point should come in contact with the wood first, to shear its portion from the solid mass, rather than force it off by the level under surface of tooth, which is a common practice. To allow the point of the tooth to touch the wood first, and thus enable it to shave its proportionate slice off by a clean cut, requires the teeth to stand hooking downward. If this position of the teeth, and their sharp points, had to be maintained altogether by the file, in opposition to their constant tendency to turn up at the point, it would involve a great expense of labor and files; but by the judicious use of a small hammer, and a steel upset with a crotched end corresponding in shape with the point of the teeth, the points may be upset and the ends brought down to the proper shape, leaving the under side of the teeth straight and true, and requiring only a slight touch with the file to complete the edge. The teeth have, in addition to the tendency to turn up at the point, another tendency equally troublesome and fully as difficult to be overcome by the file, that is, the continual narrowing of the points as they wear, which if not counteracted, makes more sett necessary in order to cut a sawdraft wide enough for clearance. The remedy for this is the same as for the other, upset the points; this not only widens and shapes the teeth, but if carefully done by an experienced hand, it consolidates and hardens the point and makes it wear better. On the other hand, if it be carelessly and bunglingly done, the ends will be bruised and slivered and broomed up so that the cure may be worse than the disease.

A sawyer who can use the little hammer and upset properly and makes a free and judicious use of the gumming machine, to remove superfluous portions and maintain the proper length and form of teeth, need spend but little time in filing. In finishing up the teeth with the file, great attention and care are required by *some persons* to get the habit of carrying the file perfectly level or at a right angle across the tooth; this is particularly essential on the under side, as a very slight bevel caused by carrying either the point or handle of the file highest in the process of filing will cause the saw to "run" towards the lowest and sharpest side of the tooth, everything else being right. The sawyer not suspecting the real cause, is tempted either to set the teeth a little more on the other side or veer the whole saw slightly to that side, to overcome the tendency to run. This setting of the saw to one side, inclines it to run to that side, while the bevel on the under side of the teeth inclines it to run to the other side. This is balancing one error by another instead of rectifying the first one, and they are constantly contending against each other, with varying success, as the kind of wood and rate of feed are varied, each having the ascendency by turns, and the saw veering slightly from one side to the other, as one tendency or the other prevails.

This condition of the saw is the cause of much of the unequally sawed lumber that is made. And it is singular how many good sawyers err in this respect, particularly when filing with the saws hung in the gate, which is the common practice with the English gate; it is still more singular how difficult it is to convince such of their error in this particular. We have frequently tried to persuade a man that he was carrying the file too high at the heel or point, and bevelling the teeth, but could not convince him without applying a square to his file and the saw, to show the variation. We have held the square to the saw to enable him to get the true level by comparison, and in three minutes after the square was removed, or as soon as he had resumed his natural gait of filing, the file would assume the old position. The only safe way for such a man is to file one half of the teeth from one side, and the other half from the other, taking particular note by trying with a square to have these alternate teeth lowest and sharpest on the outside, that is, on the side they are bent towards by the sett; this will insure equality and accuracy and the bevel, when thus distributed, is in many kinds of wood a positive advantage, enabling the saw to cut smoother and more easily.

The common recourse of the sawyer in this dilemma is, to increase the tension of the saw by straining it tighter in the gate. This will in a measure control its wavering erratic course, but it is always injurious, both to the saw and gate, to stretch it thus (like a fiddle string) and also dangerous, because a slight extra strain which is liable to occur at any time by various casualties, is likely to break the lug pins or stirrups holding the saw, or the saw itself.

We would advise any sawyer who is anxious to perfect himself in the management and control of the saw, to practise a while upon the mulley saw, which being uncontrolled by strain or other influences, shows the effect of any variation in the bevel of filing or setting most minutely, and will give him a better idea of the importance of this subject than he is likely to obtain by long practice with saws that are strained tight in a gate.

CHAPTER XIII.

SAW-MILLS.

The Mulley Saw.

THIS is an upright saw driven by a crank, and almost identical with the English gate, except that the gate or sash of the latter is omitted, the mulley being simply the naked saw running between, and kept in place by guides; the crank, pitman, and the noddle pin, having no gate and very little more than the weight of the saw to carry, are quite light. There are many different ways of connecting the saw and pitman, some of which are very ingenious, but the most common plan is to place a short piece of wood between two round or square friction rods or guides, similar to those upon which a gate runs, and attach the pitman and saw to this piece as to the under bar of a gate. We have seen mulleys working lately with the lower end guided and controlled by a working beam and rods, similar to the "parallel motion" in Watt's steam-engine, and so arranged as to give the lower end of the saw a slight jerk back just as it turns to go up, which clears the teeth from the log and leaves it perfectly free in the upward stroke. These were complicated and expensive, but might easily be simplified so as to be a very valuable improvement, as the mulley, when it can be got to take the upward stroke safely and surely without increasing the thickness of the saw, is the most economical of all saws.

The upper end of the saw has generally two small pieces of wood clamped on and running between two turned or planed guide rods similar to the lower ones. Sometimes these upper guides are only half or three-quarter inch iron rods, stretched tight, by nuts on the ends, to keep them straight. A pair of wooden gauge or guide blocks are placed just below the log, one on each side of the saw, which help to keep it in place and insure accuracy; another similar set are applied to the saw just above the log, but made so as to rise and fall, and provided either with a lever or screw to bring them either up or down, to accommodate the size of the log. The saw is thus confined to its track both above and below the log, but notwithstanding this, it is found impracticable to run saws in this way without using a thicker gauge of saws than when strained in a gate. This causes both a waste of wood and power, in cutting out the extra breadth of kerf for the thick saw, and has so far prevented the mulley from coming into more general use, and given rise to several ingenious expedients to obviate the difficulty; one of these we have already mentioned, another is to attach a strong spring to the top of the saw, which is generally done by placing a large spiral steel spring on end above the centre of the saw. The lower end of the spring is made fast to the beam upon which it stands, the upper end is bent inward to the centre of the coil composing the spring, and a half-inch rod is made fast, one end to this upper turned-in end of the spring, the other to the top of the saw, thus running through the centre of the coil, and joining it and the saw together. Now suppose the saw to be set in motion, if it was clear down when the attachment was made, then the saw would have to stretch the spring the whole length of its stroke; if it was clear up, then it would have to compress it to the same extent. This last would appear to be the best position in which to place it, as the spring, being fully compressed, strains upon the saw and is ready to exert its whole force to start the saw back at the return, and assist it in the upward stroke; but practically it is found that the full compression soon destroys the elasticity of the spring, besides the rake of the saw clears the teeth entirely from the cut at half way up, and then its momentum is sufficient, not only to make the rest of the ascent safely, but also to stretch the spring from one-third or one-fourth of its range upward. For these reasons it is found best to place the spring so as to be two-thirds compressed, and one-third stretched; in this position it not only helps to clear the saw and commence its upward stroke, but also assists to make the return at the upper range of the stroke. Any person who has noticed the effect of a spring upon a swift reciprocating motion, such as the whiffers of a carding machine, or shaking screen in a grist-mill, will easily understand this.

The best arrangement we have ever tried is the following, which answers the purpose admirably; the only peculiarity that will prevent its general application being, that it is only applicable to a double mill and inadmissible in a single one. The crank shaft must be of suitable length, and the two carriages placed side by side and parallel with each other, both cranks placed on the same shaft, that at one end being exactly up when the other is exactly down. The whole arrangement from these cranks to the top of their respective saws differs in no respect from the ordinary mulley, but instead of the spiral spring last described being attached to the top of each saw, a single wooden spring reaching across from one to the other is placed overhead; this spring is made stiff and light, and hung by its centre upon a bearing like an engine walking beam; each end of this spring is attached to the top of its respective saw by a small iron rod, and the post supporting the central bearing being made to slide either up or down, is provided with either a lever power or screw, by which the spring is raised to give the saws the proper tension or strain.

In order to get the necessary strength, and at the same time maintain the requisite lightness, the two rods connecting the saws with the ends of the spring must be joined together by a similar rod, which is carried over a short post on the centre of the spring. This spring, or walking beam it might rather be called, is of considerable length, and being elevated considerably above the top of the saws, the side swing of the connecting rods is so little that it does not disturb the top of the saws.

By using a very short crank shaft, or what would be better, a return crank, with one or both bearings outside, like the crank of a fulling mill, both saws might be made to work on the same carriage, the carriage of course would have to be made proportionately wider, and the logs both of the same length. We have never tried this, but where a deficiency of room or other reason makes the double mill inadmissible, we are satisfied that this could be done. The double mulley being so light and equally balanced in all its parts, the journals and bearings small, causing little friction, while at the same time it provides the requisite strain, to admit of the use of ordinary saws, and makes good lumber; all these considerations taken together, render it the most economical method of sawing which we have either tried or seen, as almost the whole power employed is applied to the work of cutting. When the power is very much limited, it may be still further economized by dispensing with the gig wheel and nearly all its machinery, in the following manner: Place the carriage upon rollers, or small wheels, like the large circular saw-mill carriage, and make any kind of wrag wheel that is sufficient to carry the wrag iron for the feed hands; the wrag wheel must be placed upon a wooden shaft of the ordinary form and size, but neither shaft nor carriage requires any kind of cogs or gearing, but, instead of gearing, fasten a strong rope on the cross-bar at the front end of the carriage and then wind it three times around the wrag shaft, and pass it along to the back end and wind it two or three times around the cross-bar at the end; extend the rope back to the extreme limit of the range of the carriage in that direction, and pass it over a pulley and down through the floor. If there should be sufficient height from this pulley to the bottom, all that would now be necessary would be to hang a sufficient weight upon the rope to bring back the carriage and log, but as a sufficient height is seldom found, and for other reasons, the end of the rope had better be carried back eight or ten feet further and make fast to a timber of the floor, or anything else that may be convenient at that point. This end of the rope must be fastened at such a length that when the carriage is fed forward its full extent, with the tail block up to the saw, the rope between the end fastening and the pulley over which it works behind the carriage will be stretched up almost, but not quite straight. Here upon this part of the rope between the pulley and the end, the weight must be hung, by a rolling pulley, like the weight of an eight-day clock. A considerable force is required to start the carriage at first, but after it is in motion, if the same force were continued, it would gain and come back with too much velocity. This method of fixing the rope and weight meets these conditions exactly. When the carriage is ready to be started back, the rope with the weight upon it is stretched up nearly level, and that weight in such a situation will exert a force upon the carriage several times greater than its actual weight, but after it is started, and as it progresses back, the rope slacks down more and more, and as it does so, the weight produces a correspondingly less effect, and the carriage comes to its place moderately and with little shock.

A person who never saw an arrangement of this kind working a saw carriage would think that it must feed hard, having such a weight to bring up behind it, but such is not the case: the wheels under this enable the hands to feed it along just as easily as the ordinary carriage upon the naked slides, which has to be started anew at every stroke of the saw. A carriage upon rollers like this cannot be used and fed in the ordinary way without such a weight, or a brake of some kind, because the hand gives the feeding push so suddenly and quickly that the log will be forced against the saw at every stroke, and as the feed is made during the ascent of the saw, this will spoil its operation, the ordinary carriage travelling as it does like a sleigh upon bare ground, needs no other brake.

The mulley having no gate, the feeding gear cannot

be driven from it in the usual manner. It is customary therefore to take the requisite motion from the pitman, in much the same way that it is taken from the gate in the other case, that is, by first driving a rocker from the pitman, and then from that rocker work the evener that carries the feed hands. The strokes of the mulley, however, are so very rapid, that when fed in this manner, a continuous movement is kept up, and it is preferable, when the nature of the machinery admits of it, that is, when a motion can be had from some other part of the driving machinery, to diminish a motion by belt and gear to feed the carriage forward with a smooth and equal motion. In this case a pair of cone-shaped cylinders or drums must be placed parallel with each other, and connected by a belt and a hand staff or some other convenient means provided to reach this belt, to move it towards one end or other of these cylinders, as the feed may require to be increased or diminished. These cylinders are driven by the large end of the first one, which always turns at the same velocity the feed is driven from the large end of the second one, which turns with a varying speed, according as the belt is shifted towards either end. These are not the ordinary belt cone, composed of graduated pulleys, each pair varying their relative proportions and answering the same length of belt, but they are smooth continuous tapering cylinders, placed with the small end of one to the big end of the other, and the belt fitting the same upon them at every spot in the whole length, so that it can be moved and the speed changed as little or as much as may be required, with ease and precision.

An improvement in the hanging of mulley saws has been patented, by which the saw has a swinging circular motion imparted to it as it ascends and descends, like the rocking motion given to the whip-saw by hand. This rock is given to the saw by bending the guides which direct its upper end like the segment of a large circle, which of course causes the top of the saw to describe a similar curve, and gives the face and teeth of the saw the rocking rounding motion in passing through the cut, which it is claimed enables the teeth to cut more equally, and during the whole downward stroke, and at the same time gives them more freedom in the upward stroke, and facilitates the discharge of the sawdust.

We have seen only one of these, and it running empty. Its motion is graceful, and it makes its bow both rapidly and politely, but whether its performances will justify the high superiority claimed for it by the inventor and his agents we cannot say, as we have never seen it actually at work.

The mulley saw is gaining in favor and coming into more general use. There are also more varieties, and more differences between these various mulleys in the amount of work performed, and the power applied, than in any other saw-mill, and lastly, there is still a better chance for making improvements in this than in all the other kinds of saw machines, as the mulley is yet far from having reached the state of perfection which the principle admits of, and we expect to see it so much improved, and its present well-known useful peculiarities combined and concentrated until it is the most economical and most common of all saw-mills on a small scale.

CHAPTER XIV.

SAW-MILLS.

Gangs.

WHAT is understood by "gang" in saw-mills, is a gate or sash containing a sufficient number of saws to cut up the whole log at once passing through. There are several different kinds of gangs, known by different names, according to the peculiarities of their construction, and the use to which they are put; they are all provided with some means of feeding the logs through, but as they pass through in a continuous line, and are finished at a single operation, they require no provision for returning the logs in the opposite direction.

The live gang is used mostly for logs of different qualities and sizes, and not sufficiently uniform for the other gangs. The logs are fed through this by revolving iron rollers, which instead of being fluted, are provided with large and strong spikes, upon the sharp points of which the knotty and crooked logs that are sawed in this gang are carried and held steady. They are merely split through and through in one direction, and are afterwards edged by circulars. For this purpose there are frequently two circulars placed upon the same arbor, one stationary and the other movable. The sawyer lays the board upon the table with one edge to correspond with the fixed saw, and by turning a winch at the end of the table, he moves the other saw to the proper width to edge the other side, and lets it go and picks up another to send after it as it feeds itself.

The slabbing gang and the stock gang are always used in connection, as each performs half of the work of or for the other.

The slabber has an open space left in the middle of the gate without any saws, of the same width as the stock, that is, the width the square edged lumber is intended to be, which is generally ten inches; on each side of this space enough saws are hung to cut the rest of the log into the intended thickness of lumber, and the sidings thus cut off are edged by circulars; the stock or square portion of the log is passed over on to the rollers of the stock gang to be finished.

Two iron tracks about eight inches apart run through the centre of the gate, passing through the space alluded to between the saws, and between these tracks the dogs or head blocks which hold the log travel; these dogs are not only held up by the ways, but held down, to prevent the logs from being lifted by the saws as they go up, so that the dogs can only be got in or out of the track at either end.

The dogs are three in number, each double, and holding the end of two logs at once which meet upon them; a high ledge stands up across the dog, between the end of the two logs, and strong sharp spikes run through this ledge and into both logs, the spikes being sharpened at both ends, and thus secure the logs to the dogs.

These dogs travel steadily along, holding and carrying the logs with them, and as soon as one of them arrives at the end of the ways, it is pried out from the log and carried back to the other end of the ways, and another log placed upon it when it immediately resumes its journey. These head blocks or dogs are sometimes moved forward by a large screw placed between the ways and right under the dogs, which then have three oblique cogs across their under side, at both ends, which mesh into the threads of the screw, and as it turns it screws the dogs steadily along the ways; the screw only reaches a part of the length of the ways, one log pushing another before it to make the rest of the distance.

The screw is driven by a small bevel pinion on the end, and the pinion is driven by a bevel wheel upon the wrag wheel, and the wrag wheel is pushed around by a hand in an evener, the lower end of which is hinged, and the upper end moved from a rocker as in the English gate; the different leverage of the evener adding to its power upon the wrag wheel and enabling it to turn the screw with great power. This great power is not requisite to feed the log against the saws, which if the teeth be properly shaped and sharpened, and the saws hung with a proper rake, offer but little resistance to the progress of the log: but it is required to impale the logs sufficiently firm upon the spikes or dogs of the head blocks. These cannot be driven in with the end of a bar, as they are with the English gate, because the dog is solid and the spikes are immovable, and they are inserted into the log by pressure, this is accomplished by holding the foremost head block with a bar, while the one newly placed behind is pushed forward by the screw until the dogs are pressed deep enough into the ends of the log to hold it secure. When the log comes through the saws, the sidings and slabs cut from each side are lifted, and the log, when it arrives at the proper place, is loosened from the dogs and tumbled over upon the rollers of the stock gang to be cut up and finished. The screw used for carrying these head blocks is about eight inches in diameter, and causes a great amount of friction, as the threads of the screw must move through a great distance, and travel comparatively fast, to give the head blocks the necessary speed. For this reason, the screw is now superseded to a considerable extent by an endless belt chain, made with open links, which are of an equal distance apart and pass over and hold upon knobs or projections upon a small clutch wheel or pinion placed upon the wrag wheel shaft. This chain passes over and is held extended by a similar wheel at the end of the ways, and the head blocks, instead of the oblique cogs upon their under surface to catch in the screw, have projecting toggles that catch in the open links, and attach them to the chain. The strain upon the chain in the process of dogging referred to, is so great as sometimes to break it, but if new links or portions of good chain are at hand, the delay and expense of repairing it are trifling, and it works so much more directly, and with so much less friction, that it is now generally adopted in almost all new mills. When the chain is used, the wrag wheel, feed hands, &c., are placed near the gate, a short distance from the front of the saws; when the screw is used, the wrag wheel and its accompaniments must be placed at the extremity of the ways, in either case the conveying power, whether screw or chain, must terminate at least a couple of feet before it comes to the saws. This is necessary in order to allow the foremost head block to be held back, until the one holding the hind end of the log is crowded up sufficiently to dog the log before it comes in contact with the saws. While both blocks are on the screw or chain it is impossible to dog the log, as both are held and travel at the same relative distance and speed. But when the foremost one gets beyond the reach of the chain or screw, it can be held back while the other continues on its way, and forces the dogs into the log until it is secure, when the front one is liberated and is pushed along by the log and the other block.

The Stock Gang.—The saws used in this gang are comparatively short, being only four feet, or at most four and a half feet long; these are hung at the proper distance apart for the intended thickness of the lumber, and in number sufficient to cut up the largest log that is expected to be sawn. The log or "stock" is fed forward upon three parallel fluted cast-iron rollers, eight or ten inches in diameter; these are geared together by small cog-wheels, which are so connected with the wrag wheel that it gives them all a uniform motion in the same direction, and at the same speed. Two of these rollers are placed in front of the saws, and feeding towards them; the other is on the other side, and bears up the sawed log or lumber, and carries it through. Above this fluted roller, and over the stock, a smooth roller is placed in a press frame and raised or lowered by a screw. As soon as the stock enters under this roller, it is tightened down upon it by the screw, to increase the grip of the fluted roller underneath, the weight of the log not being sufficient for that purpose. Another similar roller is placed over the stock as it feeds in, in front of the saws between the two fluted rollers, and when this one is screwed down tightly it presses the stock down upon both of these to increase the hold. In frosty weather, when the flutes become clogged with sawdust, and especially when the saws are not kept in proper order, the whole power of the screw pinching the stock between the two sets of rollers will not prevent a little slip in the feed; and for this reason short sharp spikes (like those in the live gang rollers) but smaller, are sometimes drilled into the roller next to the front of the saws.

The press frame of the roller behind the saw is mortised and keyed solidly into the ways upon which the fluted rollers rest, that in front of the saws would be in the way of the sawyer in hanging the saws or pointing the teeth, and it is either hinged to the ways with strong iron clamps, or hung by a roller overhead, in either case, admitting of being swung back out of the way, and easily straightened back into its place.

The fluted rollers are supplemented out to each end of the ways with smooth ones, generally of wood, to complete the support of the stock both ways. All these gangs are frequently fitted to take through two logs at once, and when the grade of logs is very small, they sometimes pass three logs abreast.

"The Yankee gang" is a compound of the last two; the saw gate being double, containing a gang of long saws for slabbing on one side and a gang of short stock saws on the other, with an upright stile between, making a division in the middle; the ways are also double, having a set of head blocks carried by a screw or chain, on the slabbing side, and a set of rollers upon the other side. These are each complete in its details like the slabber and stock gang, but combined in the same gate and carried by the same power.

The logs are generally drawn in and placed upon the slabber head blocks at the up-stream end of the ways, then passed through the slabbing side of the gate and canted on to the stock rollers, when they pass back in the opposite direction through the stock saws in the other side of the gate and are finished; the slabbing saws are thus facing up stream and the stock saws down stream, with a steady stream of logs passing through the gate in both directions, when it is at work.

The Yankee gang requires no further description, as each side is exactly like its single counterpart already described; and its peculiarity consists in combining both in the same gate to be carried by the same power. All these gang gates are made in the same way, the cross-bars being double, with a space between for the stirrup irons, that hold the saws, to pass through.

Each saw is drilled through both ends, and a lug pin about half an inch in diameter, and three-quarters long, is fastened into the hole by riveting. The stirrup irons are made with a slit in the end to admit the end of the saw, each side being turned in a hook-shape to catch upon each end of the lug pin; the lower stirrups have a cross head that is slipped down between the bars, and then turned crosswise, when they hold upon both bars alike. The upper ones have a key way made through the end about two inches long, and wide enough to admit a key made of saw plate. The keys are cut about ten inches long, one and a quarter inches wide at the point, and two and one-fourth at the head; these keys are driven through the head of the upper stirrups to strain the saws, holding upon the cross-bars of the gate.

The saws must be arranged and kept in place by some kind of gauges. These consist of either blocks of wood of equal and proper thickness placed between each saw at both ends, or iron gauges reaching across all the saws, one behind and another in front, at both ends. The iron gauges have deep slits cut into them by machinery, of the right width to slip into the saw plate, and graduated to the proper distance apart, for the intended thickness of the lumber, that is, the thickness of the stuff and about one-twelfth part of an inch additional for set or clearance on the saw draft. These gauges are made of broad flat bar iron, and have a gauge for different thicknesses cut in each edge, so that each piece of gauge will answer for two different thicknesses of stuff, by merely turning it over.

A crotched iron is bolted on to the gate stile at each

end of the gauge, to which they are fastened, by the ends passing through the slit or crotch, and these ends being rounded and screwed, have each two nuts upon them, one on each side of the crotch. By turning these nuts, the gauges are shifted until the saws are brought plumb and true, and are then fastened by tightening the nut against the crotches. Only the two gauges holding the backs of the saws require this provision for shifting, those holding the fronts are bolted flat on the two cross-bars of the gate, and being placed carefully and true at first, they do not require to be shifted. These gauges are expensive, as the iron of which they are made must be pretty thick to be safe against accidents, to which they are continually liable, by the sidings or slabs getting loose or split and coming in contact with them. This thickness makes the cutting of the recesses for the saws a tedious process, and a complete set of this kind of gauges for all the various thicknesses cut for both slabber and stock gang, would amount to quite a sum, and it is customary to have these only for some of the most common thicknesses, and then use wooden gauges for all other grades of lumber.

The wooden gauges or blocks between the saws answer the purpose when once in and tightened up all right; but the great objection to their general use is, that they require too much time to change the set of saws, and thus furnish the constant temptation to run the saws too long without changing or sharpening.

The saws in a gang are so close together that they cannot be filed in their place like a single saw; besides, there are so many of them that it would cause too much waste of time to have the mill idle and the rest of the hands waiting until the saws are filed; hence it is customary to have two sets of saws, one set being filed and fitted while the other set is at work. When provided with these iron gauges, an expert hand will change a gang in twenty or thirty minutes, or while the other hands are at their meals, and then it is customary to change two or three times in the twenty-four hours, but when the wooden blocks are used for the gauges, it takes several hours to make the change, and the sawyer will point the teeth, once at least, and perhaps two or three times in the gate to get the saws to run as long as possible without changing.

Many years ago we used a very simple iron gauge which answered the purpose well, and the same gauge could be adjusted to any thickness. Why it has never come into use, is a question for solution. It was made of a piece of round iron of the proper length, screwed from end to end the whole length; this was fastened in two crotch irons at the end like the other, and the saws had a slit cut in them by drilling a hole at the proper place and cutting the edge through to it with the gummer. The screwed rod passed along in this slit, and a nut on each side of each saw was adjusted to the required gauge or thickness, which secured the saws in place. A wooden gauge or pattern was required, with the saws and spaces marked upon it, for expedition and accuracy in setting these for any particular thickness of stuff, but when once set, there was no further trouble with them more than the other kind, until a different thickness of lumber was required to be sawn, when the nuts were re-adjusted by the pattern representing the grade wanted.

The wooden gauge blocks referred to are made of cherry, or some other good and moderately hard wood, about six inches long and three and a half or four inches wide, and one-twelfth of an inch thicker than the stuff intended to be sawn.

A piece is cut out of both sides at the middle to clear the stirrups, and a hole bored through, opposite the lug-pins. These clearances are required, to let the blocks come in contact with the saws. One corner is also bevelled off from each piece, from about the middle of the end to the middle of the edge, of all the upper blocks, for clearance and convenience in placing and removal. To hang the saws with these, the outside saw is placed and adjusted by four set screws, that are screwed into a thick plate sunk into the stile of the gate; the heads of these bolts are square for convenience in turning, and stand out so as to come in contact with the saw, or rather with a block that intervenes between the bolt and saw, when the first one is hung in. These bolts, two above and two below, are screwed in or out until that saw is made plumb and true, then a pair of gauge blocks are placed, and another saw alternately, until all are hung in place, when the last one is secured by the four set screws and intervening blocks, the same as on the other side; each saw is then strained or tightened separately by driving the key through the head of the upper stirrup, as already indicated. The stirrups used for the upper ends of the saws are, so far as we have seen them, nearly all alike, but there are different ways of attaching the lower end of the saws to the bars of the gate, one at least of which is more convenient than the long loose stirrups we have described. This consists of a piece of strong iron, like boiler plate, with the upper edge curled over in the shape of a hook, and then bolted on to the gate bar; this makes one continuous hook from end to end, on any part of which there is a good grip and free chance to shift either way. To match this, the saws have each a short hook stirrup, fastened permanently on the lower end, and this is slipped down and hooked into the hooking edge of the boiler plate, and is easily adjusted to the gauges when they are applied.

A gang gate when made of wood must be of good timber and well put together; it must be ironed along all the bars where the stirrups have their bearings, besides being well bolted and clamped together at the corners and all the principal joints; otherwise the straining of so many saws in it with the jamming and banging it has to endure, independently of accidents, soon wears it out, especially if the sawyers are principally rough green hands. Although such a gate is considerably the cheapest at first, an iron gate is certainly the cheapest in the end. An iron gate is simple in construction, the main item of expense being in the weight of the material, and is not affected by wear and tear like the wooden ones. A good one will wear during the life or running of an ordinary mill, and then be worth nearly as much as at first, to put into another mill.

All the details of the gang machinery, aside from the gate and gig-back works, are identical with those of the English gate, only on a larger scale. The power of wheel, and strength of shafting, crank, pitman, &c., are about double those of the English gate. The crank used, however, is shorter in the stroke than that of the English gate, being from eight to eleven inches, while the other is from eleven to fifteen; this gives it more purchase on the pitman and gate, and helps to account for its carrying so many more saws and so much more weight, at nearly the same speed, as the English gate carries its single saw, and all this with only double the power of the other.

The gang, taken altogether, is the most economical method yet devised for making lumber on a large scale.

The saws used are thinner than any others, either upright or circular, the difference being sufficient to average at least one piece more lumber out of each log. The saving of this extra piece of lumber is a consideration, but scarcely as much as saving the power and wear, and tear that are otherwise required to cut that extra piece into sawdust. Again, all the time and labor consumed in gigging back and setting the log over for the other kinds of saws is saved, with the power and wear and tear that that process requires. The hands attending are also more economically employed, having only to supply and feed the stream of logs passing through and carry away the lumber. When we add to these considerations the fact that only double the amount of power is generally applied to the gang carrying two or three dozen saws that is applied to the English gate with a single one, and little more than half the power generally applied to the large circular, the economy of the gang will be apparent. In addition to the economy of the gang in power and timber, we have another point worthy of regard, namely, the superior quality of the lumber made. Gang sawn lumber, as a general rule, is truer, more equal, and smoother, than that sawn by either the English gate, mulley, or the big circular, the difference being sufficient to average one cent in value upon each piece in favor of that sawn by the gang. This preference is partly due to the gang cutting completely through and leaving no stubshot end.

The big circular of course cuts through the same way; but, on the other hand, the lumber made by it is the roughest and hardest to dress or plane, as it is always more or less circled and ridged by the variations of the saw in its revolutions.

The form and construction of the wooden gang gate will be understood by the following description: The upright stiles may be of spruce, it being stiff and light, about six inches by seven, the cross-bars of good hard wood, three inches thick and about a foot wide: these three-inch bars are notched into the seven-inch stiles, one on each side, and made even and flush outside, which will leave an inch space between for the saw stirrups. The size must be determined by circumstances, that is, the height must be made to answer the length of saws intended to be used, and the width by the number of saws intended to be hung in it, or, rather, by the general size of the logs to be sawn in it, and whether intended for a single log or two abreast.

The Yankee gang has a third upright stile dividing it in the middle, and two short bars mortised in below the upper cross-bar, to shorten the space for the stock saws. The common width for each side of the Yankee is thirty inches clear space, the centre dividing stile being the same width as the others, but only three and a half or four inches thick.

CHAPTER XV.

THE CIRCULAR SAW-MILL.

THE large circular saw used for cutting ordinary round saw logs into lumber, is from forty inches up to six feet in diameter, according to the size of logs intended to be worked up. The saw is hung upon a horizontal arbor, and generally made of cast steel, about three inches in diameter, and six or seven feet long. This arbor, as well as the feed and gig-back shafts, is hung in a wooden or iron frame, about four feet wide and seven long, and from nine to twelve inches deep. This central frame is made of four, or sometimes six pieces, fastened at the corners with bolts, so that it can be taken to pieces, to ship or move from one place to another. The collars, securing the saw on the arbor, are generally of wrought iron, the movable one being tightened against the saw with a screw nut. The arbor runs in adjustable babbit boxes, both being sometimes cast in a movable piece which reaches across the main frame, and can be shifted to give the saw more or less lead.

The feeding works consist of a cone of three pulleys, for a three-inch belt, placed upon the saw arbor; a similar cone is placed upon a small shaft lying parallel with the arbor, the two cones being connected by a belt. On the other end of this small shaft is a small leather pulley three inches in diameter; under this last shaft is a third one, with a small pinion on the end, which reaches under the carriage, and gears into its rack of cogs. Directly under the small leather pulley, and on the last-named shaft, is a large iron pulley, turned true; along side of this is another pulley of the same size, with flanges to keep on a belt which hangs loosely around this and a small one on the saw arbor. These are controlled by two levers; one presses the small leather pulley in contact with the large iron feeding pulley, thus feeding the carriage forward; the other presses a tightener against the slack belt, and thus gigs the carriage back. The two levers are placed so that the sawyer can both feed and gig back without changing his position.

The head blocks are of iron, and the more modern are shifted and set by screws; experience having proved the screw to be the most reliable and accurate device for that purpose.

The carriage is made in sections, for convenience in transportation; it is set upon small wheels which run upon an iron track, and is generally made to cut stuff twenty-four feet long. By lengthening the carriage and track it can be made to cut any reasonable length.

A portion of the centre of every circular saw is covered by the collar which secures it upon the arbor, and it is only the depth from this collar to the circumference of the saw that is available for the cut; therefore a large diameter of saw is required to cut very large logs. This necessary increase of size causes a large increase of original cost, in the first place, and it also involves a corresponding increase of thickness in the saw, which increases the width of kerf, and consequently waste of wood and driving power, in like proportion. These limit the economic application of the circular saw to about five feet diameter; and when deeper cutting than this size will reach is required, it is found to be more economical to use two saws, one of four feet or less, placed in the ordinary position under the log, and a smaller one on an arbor above, and a little behind, and parallel with the lower saw, to complete the cut.

The sizes of these two saws should be graduated by calculating the probable average sizes of the logs to be sawn; the lower one should be large enough to cut all the medium and small-sized logs, and the upper auxiliary one only large enough to complete the cut of the lower one through the larger logs. By this arrangement a thinner and cheaper gauge of saws is used, with less waste of timber and driving power, and less danger of kinking or buckling the saws. And as the greater part of the work is performed by the lower saw, while the upper one remains idle, they are also much easier kept in order than the very large single saw which would be required to do the work. For these reasons the double saws are much used in California and the Western States, and

Fig. 36.

Cooper's Circular Saw-Mill.

(Independent and simultaneous screwhead blocks.)

Manufactured at the Mount Vernon Iron Works, Mount Vernon, Ohio.

wherever the growth of timber is large; while in the Northern and Eastern States and adjoining British Provinces, where the timber is lighter, the single large circular is much oftener used.

It would be a useless occupation of time and space to enter into a fuller or more minute description of these machines, as they are now made in regular machine shops by experienced workmen, who understand all the details of their parts and construction much better than we do, and as they are all completed and ready to set to work, the ordinary millwright or sawyer has nothing to do to one when it arrives from the shop but to put it properly together and run it. We will, therefore, only give such instructions as will enable the millwright or sawyer to do this.

First place the ways for the iron track perfectly raight and level, parallel with each other, and out of wind; then strike two chalk lines upon each way, the width of the iron track apart, for a guide by which to place the pieces of track. Place the track to balance equally between these two lines, but before fastening it down, stretch a line the whole length along the guiding corner of the track, if it be a square one; if V shaped, along the apex, and regulate any inequalities in the different portions by this line. Next place the sides of the carriage in position on the track, and then place the centre frame carrying the saw-arbor and machinery upon these, taking particular care to have the track and the arbor at right angles with each other. See that the arbor turns free in its boxes, without being loose. The saw collar makes a shoulder against the box, to prevent the arbor from shifting endways in that direction; a movable collar at the other side of the box forms the other shoulder; this should be placed about one-twentieth of an inch clear of the box, to admit of end play, and made fast by its set screw. The nut and loose collar may now be taken off, and the saw tried on; it should be a little slack, both on the arbor and bolts, otherwise if the arbor should heat it might injure the saw. The loose collar may now be put on and tightened up by the nut; see that the collars fit perfectly true and even upon the saw; the fast collar should be true next the saw, but the loose one may be slightly concave. If they do not embrace the saw equally, rings of paper should be interposed to perfect the bearing, and the nut again tightened, now hold a gauge firm upon the saw guide against the side of the saw, and turn it round by the pulley and see if it runs true; if not, slips of paper must be put between the collar and saw at the point to which it inclines, until it does run true. When thus adjusted, to train true by the gauge all around, take hold of the saw and jerk it back to make it grip tight on the driving bolts, and tighten up the nut. It will facilitate the hanging of the saw at any future time, to mark it and the fast collar, so that the same side of each can be placed together every time it is rehung.

The saw should be perfectly straight and true on the side next the log, all the dish or convexity being on the slab side; you should test this by a straight edge before attempting to try it in a log. We have known several sawyers to be perplexed by getting the saw with the dish to the wrong side, when everything else about the machine was correct; and they were trying in vain to find the fault in something else.

The metallic disk in front of the saw, which carries the board, also springs it out from the cut, and relieves the saw on the slab side. This gives it a slight tendency to run towards that side, and to counterbalance this, the saw must be set (hung) with a slight inclination towards the log. To do this, run the carriage up until a head-block is opposite the cutting side of the saw, measure the distance exactly from this side to the block, then move the head-block forward to the back side of the saw and measure the distance there. The distance at this side should be about the twentieth part of an inch more than at the cutting side; this is the average difference, but circumstances may require this to be varied, some requiring more and some less.

These saws should be geared so that the teeth will travel at the rate of about seven thousand feet per minute, when cutting; and to obtain this working velocity without an excessive and dangerous speed when running empty, the water-gate should be regulated by a governor, to control the speed, the same as when driven by steam. The want of this is the most common defect in circular saw-mills, when driven by water. A circular saw-mill when driven by water, and maintaining anything near a uniformity of motion when working, and running idle, is very rarely seen; in fact a great proportion of them check up very perceptibly in running through an ordinary cut, and many will almost or quite choke in a deep hard cut, requiring the feed to be stopped and perhaps the log slacked back, to allow it to gain speed and momentum sufficient to carry it through the cut.

This is a great source of trouble to the sawyer, and injurious to the saw and machinery, and when a man has gone to the expense of starting a circular saw-mill thus defective, the best policy he can pursue is, to lay out enough more upon it to furnish a sufficient power and some convenient method to control and regulate the power to the varying requirements of the machine. The large and unwieldy gate supplying the water to most of the wheels used for driving the circular saw-mill does not admit of control by a governor; but several of the wheels used admit of an additional chute and gate, and in this case the original gate should be opened to its full extent, and the small additional one regulated to furnish or withhold the auxiliary power, as required. When this plan is not admissible, then a small additional wheel may be put in, and the governor attached to its gate, for it is worth considerable trouble and expense to get rid of such a nuisance, especially where the water is limited in quantity or duration, for the water must be economized in order to cut out the quantity intended, while on a large and durable stream the sawyer may balk and bungle with a mill defective in power, and yet cut out a good stock of logs. Many good millwrights err in thus furnishing too little power to the circular, expecting to economize the water, but it is mistaken economy; it is like sending a small team to transport a heavy load over a long journey—they will stick in every mud-hole, and balk at every hill, while a team a few hundred pounds heavier would walk comfortably along with the same load.

The power required to drive the circular saw depends on its size and the quantity of lumber it is intended to cut. Thus, twelve horse power will drive a forty inch saw, and cut about 2500 feet of soft wood lumber in twelve hours; fifteen horse power will drive a four foot saw, and cut about 4000 feet in the same time; twenty horse power will drive a four and a half foot saw, and cut from 6000 to 7000 feet in twelve hours. Twenty-five horse power will drive a five foot saw, and cut 8000 to 9000 feet of soft lumber per day. Thirty horse power will drive a six foot saw, and cut from 9000 to 11,000 feet per day. Forty horse power will run the largest saws that are made, and cut from 12,000 to 18,000 feet per day. This is about the average amount of soft wood lumber that each will cut when in ordinary condition, in a day of twelve hours, including all ordinary stoppages.

In order to ascertain how many horse power any particular head and quantity of water will furnish, see Tables in another part of this work.

We neglected to give in the proper place the necessary directions for setting and sharpening circular saws. The saw is set by bending each alternate tooth to one side, about one-sixteenth of an inch; the intermediate teeth being of course bent to the opposite side the same distance, the amount of set must be varied according to the kind of wood and other circumstances. The most convenient set is an iron about ten inches long, with a slit of the proper depth and width, cut out of the end, the bottom or termination of the slit being drilled out, or widened to save the corners of the teeth by griping clear of the point. This should be bent to a right angle near the slit end, so that the strain of bending the tooth will be applied in the direction of the saw plate, instead of sideways. A very hard saw may be set more safely by holding the outside of the tooth against a suitable iron, and striking the other side near the point, with a hammer. Some kind of gauge should be used, to insure equality in the set, particularly with circulars, as these cannot be accurately tested either by the eye or a straight-edge. A good gauge is made of a thin piece of hard wood cut in the shape of a coffin, about six inches long, and three wide; four small wood screws are inserted into this, for legs, the two at the shoulders and the one at the foot being exactly level, but the one at the head say one-sixteenth of an inch lower (shorter) than the others. Place this gauge with the first three nail-heads against the side of the saw, and the fourth short one in a line with the point of the tooth, and adjust all the alternate teeth on both sides of the saw to the short nail-head.

To true the points of the teeth, hold a piece of stone firmly on the saw guide, and let the saw run about half speed, move the stone carefully forward until the longest teeth are ground off, and the shortest one touches the stone; then file all the points or under sides of the teeth first, and, in doing this, the caution given for filing the under side of upright saw teeth must be strictly attended to, as a slight bevel to the cutting side of the teeth of a circular will affect its running as much as a mulley saw.

The arbor in this case furnishes a constant source of comparison by which to regulate the bevel of the file; then file the backs of the teeth down until the ground points are just visible.

The teeth of the circular have the same tendency to wear off in front, and narrow at the point, as upright saw teeth, and require the upset punch and little hammer to be used freely; the latter in the hand of an experienced sawyer, with a six square steel mandrel to hold against the under side of the tooth, is the best. This mandrel, if eight or ten inches long, and tapering from two inches in diameter at one end, to three-fourths of an inch at the other, will answer for any pitch of teeth. See remarks on this subject in the article on upright saws.

The Edging Circular.

The Table for the edging saw is from 24 to 30 feet long: 2½ to 3 feet high, and about 2½ feet wide. It is made with two parallel sticks, four inches thick and six or eight deep, supported upon legs of the same material at suitable distances.

The box or bearing of the arbor next the saw is set into the back stick of this frame, the saw running over it into the inside.

A short counter frame is extended out from the centre of the main one, supported on two extra legs, into which the other box of the arbor is set. The arbor, with its pulley and driving belt, is thus outside of the main frame, leaving the space between the parallel sticks clear for the carriage and its supporting rollers or tramways.

Sometimes parallel wooden rollers are hung upon iron journals between these sticks for the carriage to traverse upon. Sometimes iron wheels are placed, two upon each iron journal or shaft, instead of the wooden rollers; in either case, the table must be furnished with these at suitable distances, its whole length. A preferable way, because more accurate and durable, is to place four small cast-iron wheels under each side of the carriage, and either a square or V-shaped track under the wheels, the whole length of the table; in this case a piece should either be rabbeted out of the upper inside corners of the two sticks composing the frame, or a piece spiked on to the inside of each to place the iron track upon, as the carriage would be too high above the table if the track were placed on top.

When the carriage is completed and located on its track, a board or plank must be placed along the back stick of the frame to level it up to the surface of the carriage, or rather one-eighth of an inch lower; this is required to support the edgings or strips cut off by the saw, which are behind it, and consequently over the back edge of the carriage. When a circular is to be used for splitting wide stuff down the middle, the saw is sometimes made to run in a slit in the middle of the carriage. In this case the parallel wooden rollers referred to above are the best for transverses, as they support the carriage equally along its whole width, and this it requires, because the slit in which the saw works divides the carriage in two, and prevent any cross connection except at the ends. The wooden rollers furnish the necessary support to the middle of the carriage, which otherwise would bulge down by its own weight, and that of the stuff to be sawn.

It is the practice of many mills merely to straighten both edges of the stuff as wide as the piece will make, one edge being first cut off, and the piece turned over and run through again, to straighten the other edge; this saves all the square part in one piece, but without any regard to width, or having the two ends equal.

At other mills, the stuff is slit down to a certain width and parallel, the surplus pieces cut off being worked up by other saws into fencing, strips, laths, &c. When this is done by a single edger, two gauge-pins are set into an iron plate, one near each end of the carriage, and sunk down in a groove in which they can be moved and set by a marked scale to any required width.

In some large mills two saws are hung side by side, one being secured on the arbor in the usual way, the other movable and made to slide closer to, or farther from the fast one. This is moved by a handle projecting from the end or side of the table, and set each time to suit the varying width of stuff: this is generally geared to feed itself, and of course edges both sides at once, both edges being made parallel.

The same rules for hanging, setting, and sharpening the large circular, will apply to the edger, shingle, or lath saws, and need not be repeated, except that circulars for cross-cutting should be filed with more or less bevel according to the kind of wood and work, and the forward hook of the teeth dispensed with; the front pointing to the arbor, instead of to a circle two-thirds of the diameter of the saw, which is the best rule for the front of the slitting teeth. In some circumstances the teeth of a cross-cut circular may be slanted equally from both sides like the teeth of a hand cross-cut, and will work safer and better, it then crowds the wood slightly away from it, which tends to regulate the feed, or rather admits of greater freedom and safety in feeding. The velocity at which the edging circular should be run depends upon the size of the saw, the amount of power, and the work intended; in ordinary circumstances a fifteen inch edger should make from fifteen to eighteen hundred revolutions per minute. It is desirable that the edger, like the circular, the barley millstone, smut mill, and all other swift revolving machines, should be geared to increase the speed as little as possible when running empty.

Log Table.

We give below a Table of Logs reduced to inch board measure, showing the amount of one-inch stuff that can be cut out of logs of various lengths, from ten to twenty-five feet, and of different diameters from twelve up to forty-four inches.

To find the amount of lumber a log will make: First find the length of the log in the first, or left-hand column; then on the top of the page to the right, find the diameter, and under the diameter, opposite the length, will be found the quantity that length and diameter of log will make.

Length in feet.	DIAMETER IN INCHES.																
	12	13	14	15	16	17	18	19	20	21	22	23	24	25	26	27	28
10	49	61	72	89	99	116	133	150	175	190	209	235	252	287	313	342	363
11	54	67	79	98	109	127	147	165	192	209	230	259	278	315	344	377	400
12	59	73	86	107	119	139	160	180	210	228	251	283	303	344	375	411	436
13	64	79	93	116	129	150	173	195	227	247	272	306	328	373	408	445	473
14	69	85	100	125	139	162	187	210	245	266	292	330	353	401	439	479	509
15	74	91	107	134	149	173	200	225	262	285	313	353	379	430	469	514	545
16	79	97	114	142	159	185	213	240	280	304	334	377	404	459	500	548	582
17	84	103	122	151	168	196	227	255	297	323	355	400	429	487	531	582	618
18	89	109	129	160	178	208	240	270	315	342	376	424	454	516	562	616	654
19	93	116	136	169	188	219	253	285	332	361	397	447	480	545	594	650	692
20	98	122	143	178	198	232	267	300	350	380	418	470	505	573	625	684	728
21	103	128	150	187	208	243	280	315	368	399	439	495	530	603	656	719	764
22	108	134	157	196	218	255	293	330	385	418	460	518	555	631	688	753	800
23	113	140	164	205	228	266	307	345	403	437	480	542	571	659	719	787	837
24	118	146	172	214	238	278	320	360	420	456	501	566	606	688	750	821	873
25	123	152	179	223	248	289	333	375	438	475	522	589	631	717	781	856	910

Length in feet.	DIAMETER IN INCHES.															
	29	30	31	32	33	34	35	36	37	38	39	40	41	42	43	44
10	381	411	444	460	490	500	547	577	644	669	700	752	795	840	872	925
11	419	451	488	506	539	550	602	634	708	734	770	828	874	924	959	1017
12	457	493	532	552	588	600	657	692	772	801	840	903	954	1007	1046	1110
13	495	534	576	598	637	650	712	750	836	868	910	998	1033	1091	1135	1203
14	533	575	622	644	686	700	766	807	901	934	980	1053	1113	1175	1222	1295
15	571	616	666	690	735	750	821	865	965	1001	1050	1129	1192	1259	1309	1388
16	609	657	710	736	784	800	876	923	1029	1068	1120	1204	1272	1343	1396	1480
17	647	698	755	782	833	850	931	980	1094	1134	1190	1279	1351	1427	1484	1573
18	685	739	799	828	882	900	985	1038	1158	1201	1260	1354	1431	1511	1571	1665
19	723	780	843	874	931	950	1040	1096	1222	1268	1330	1430	1510	1595	1658	1758
20	761	821	888	920	980	1000	1095	1152	1287	1335	1400	1505	1590	1679	1745	1850
21	800	863	932	966	1029	1050	1150	1210								
22	838	904	976	1012	1078	1100	1204	1268								
23	876	945	1021	1058	1127	1150	1259	1322								
24	914	986	1065	1104	1176	1200	1314	1380								
25	952	1027	1109	1150	1225	1250	1369	1438								

CHAPTER XVI.

GRIST-MILLS.

THERE is no rule that can be set down as of general application for the construction of the building for a grist-mill. Each new site selected to build upon, the millwright will find different in some respects from every one he has ever before built upon; and there are so many contingencies to be considered, and harmonized, that he will have full scope for all the ingenuity and common sense he may possess, to reconcile these conflicting considerations, and determine all the points involved, to the best advantage. First, are the height and features of the bank, then the height, entrance, and exit of the water; the road or approach, for the convenient delivery and loading of the grist, next the kind and quantity of work, or grinding intended to be done; and the kind, quantity, and position of the machinery required to do that amount of work, with this particular head and quantity of water; lastly, the material of which it is to be composed.

Here we may remark that whatever material may be selected for the main structure, the foundation should, if practicable, be of stone at least as high as it will be exposed to water, and the height will depend in a great measure upon the kind of water-wheel used. With any of the different turbine wheels (wheels upon a vertical shaft) a low basement, barely sufficient to work conveniently in, will suffice; as the flume can be carried up through the lower floor of the first story, containing the bridge trees and machinery; or it may be outside, or in a detached space and the water carried under this floor to supply the wheels.

When the overshot wheel is used, and placed inside the mill, the stonework should extend at least to the height of the wheel, and the spout or flume supplying it with water, and the stones, bolts, &c., should if possible be placed above this level, because when the dust accumulates in contact with the spray, or dampness from the water, a fermentation ensues which soon affects woodwork and everything else susceptible of decay. But this arrangement might clash with another consideration equally important in the construction of a custom mill, which is, that the main floor occupied by the stones, &c., should be at the proper height to correspond with the road and mill-yard, for the convenient loading and unloading of teams at the door; this is best accomplished by extending a platform around the door at a convenient height for a wagon, and this platform should agree with the mill floor, so that a truck or other wheel contrivance may be used for carrying the bags.

We have geared several mills lately, with from fourteen to thirty feet head of water, the water being conveyed along the bank behind, in a canal, and the buildings on a flat below, with the mill-yard and road only a few feet above the level of the tail-race; some of these were driven by overshot wheels, varying from fourteen to twenty-seven and a half feet, and others by turbines of various kinds. When turbines were used the flumes were always placed behind and outside the mill, the water being conveyed to the wheels through the openings left in the foundation walls for that purpose. When overshots were used, they were placed in a separate wheel-house, also behind, and between the mill and the bank; these were geared by placing spur geared segments around the outside of the rim of the water wheel next to the mill, with a pinion working in these; this pinion was placed on the descending side of the wheel and near the bottom; a shaft from this pinion was carried through an opening in the foundation wall to the inside of the mill where a bevel wheel was placed upon it. This wheel geared into a bevel pinion on an upright shaft which carried either a spur wheel or belt-drum, from which the spindles and stones were driven. A small shaft was coupled on to the top of the upright shaft and carried up through the mill, to drive the bolts, elevators, &c.

This upright movement is sufficient for two run of stones, or three at most; where more are intended, the horizontal shaft is extended further into the mill and another set of bevel gearing, upright shaft, &c., similar to the first is set up, and the work divided between the two.

We have also geared some overshot mills where the road and mill-yard were on the bank, and the wheels inside of the mill, and underneath the machinery; two of these are belt-mills and were geared in the following manner: A bevel gearing was put on, the whole size of the water wheel, by bolting segments around the outer face of the rim holding the buckets; to do this we placed an extra thickness of plank around that rim, four inches larger in diameter than the original rims, and thus projecting two inches past the wheel all around. In this was turned out a groove or recess the exact size and shape of the segments, the outside of the groove being equal to the circumference of the wheel; this was necessary in order to get a substantial shoulder bearing in the wood for the segments (without which, the bolts alone would not last long) and at the same time allow the bolts to pass through near the outer edge of the rim to clear the buckets. A bevel pinion matching into these segments, was placed upon an upright shaft resting on a bridge tree along side of the water wheel, and a drum built upon this shaft carried the belts for three run of stones and a smut machine; a small shaft was coupled on to the drum-shaft, as in the others, to drive the elevators, &c.; each belt had a bearer under it to keep it in place when slack and idle. These bearers were let into a groove turned in the drum for that purpose, and thence extended to the sheaves upon the spindles and around under these. A movable tightener was hung between two horizontal arms in an upright shaft, standing outside of each belt, on the slack side between the drum and sheave; these were brought against the belt by a weight attached to a rope passing over a pulley, and the other end made fast to the tightener frame. By this means the belt was tightened and the spindle and stone set in motion; by withdrawing the tightener and raising the weight, the belt was slackened and rested upon the bearer, and the stone was stopped; another rope was attached to the opposite side of the tightener frame, and passing up through the floor was connected with a windlass or handstaff convenient to the stone. Thus the tightener is controlled, and a run of stones can be started or stopped, while the rest of the machinery is in full motion.

Another mill, driven by a twenty foot overshot, was geared in the following manner: An extra rim was put around the arms seventeen feet in diameter, and bevel gearing put upon this; a crown wheel or pinion four feet in diameter to match this gearing was put upon an upright shaft, and upon this shaft a spur wheel, nine feet and five inches in diameter, was placed, working above the water wheel, and gearing into pinions of twenty-two inches in diameter on each spindle. These pinions are thrown out of gear by a long lever working under each, a hand-staff attached to the long end reaching up through the floor for convenience in throwing them out or in gear. This requires the whole machinery to be stopped, either to stop or start a run of stones, and the miller has generally to go below and shift the pinion a little that it may enter the cogs before it can drop into gear. Another objection to this arrangement is, that the miller, anxious to save time, will sometimes drop the pinion before the spur wheel is entirely at rest, and break the corners off the cogs.

The stonework of this mill is twenty-four feet high at the end where the wheel is situated, the other end running upon the bank where the wall is only three or four feet above the surface. A wooden (frame) building is set upon this, in which the stones, bolts, &c., are placed.

Another overshot mill is geared in the following manner: the wheel is twenty feet in diameter with three rims, and two tiers of five feet buckets. It is placed close to the end wall, and around the rim next the interior is a spur gear of heavy segments; at the descending centre of these segments a shaft is placed, with a pinion on one end gearing into these, and a bevel wheel on the other, this bevel wheel drives a bevel pinion and upright shaft carrying a spur wheel, seven feet in diameter. This spur wheel drives two pinions secured upon two spindles, each spindle being supported upon its respective bridge-tree under the stones; these spindles, however, do not carry their mill-stone upon a bail and driver in the ordinary way, but terminate in a true turned cone at the top. This cone is surmounted by another concave portion, or hollow cone, which fits it exactly, and is attached to the running mill-stone.

To start these stones, the pinion and lower or conical portion of the spindle being in motion, the bridge-tree supporting these is raised, thus bringing the cone and socket in contact, and driving the mill-stone by friction; to stop the stone, let down the cone until out of contact, and the stone will remain at rest, while the cone continues to revolve with the rest of the machinery. On the other, or ascending side of the wheel, and at the same height (the centre), is another pinion working in the segments, and having its shaft, bevel gearing, and everything else, exactly like the other set, and driving two more run of mill-stones in the same manner. At the highest point of the wheel above is a third pinion, working in the same segments, and driving an upright shaft which carries the smut machine, and all the machinery above. This appears to be a substantial and durable combination of machinery, and provides for stopping or starting a run of stones, with the other machinery in motion, as conveniently as the belt gearing, and when all the stones and machinery are in operation, it answers the purpose well; but when only a part, or say one run of stones is working, all the other pinions, and shafts and wheels, are running empty and idle, and this involves a considerable waste of power, and useless wear and tear of machinery.

We intended ending our observations on the different plans of constructing these overshot mills with this one, but since there is such a variety of these it may contribute something to assist in determining the most suitable arrangement for a new mill. We will describe a few more variations in detail, and we do this the more willingly, from a consideration of the fact, that much of the difference in mills occurs among these overshots, those driven by the various wheels of modern invention admitting of much more uniformity of construction.

The following method of belt gearing was introduced several years ago by Columbus Smith, a celebrated American millwright. He geared several mills with wheels varying from sixteen to twenty-four feet in diameter, each having three rims with two tiers of buckets, varying from five to eight feet in length, and driving from three to six run of stones each. These wheels had a spur gearing of segments placed around one rim, with a pinion to match, placed about the centre of the descending side; this pinion drove a long heavy shaft extending through the lower story of the mill, and parallel with the husk-timbers, and bridge-trees; upon this shaft a long continuous drum (or several short ones) was placed, about eight feet in diameter; a bevel pinion was put on each spindle, and another bevel pinion gearing into these was placed upon a short horizontal shaft at the foot of each spindle, and lying parallel with the large drum. Upon the short shaft a sheave or pulley was placed along side of the bevel wheel and a belt passing around this pulley and the big drum connected them together; the driving side of this belt was below, and a heavy tightener was placed upon the upper, or slack belt, to start or stop a run of stones.

To effect this, a rope was made fast to the frame of the tightener, and the other end attached to a windlass above, near the stones. By slacking down the rope by the windlass the weight of the tightener upon the belt caused it to gripe and start the pulley, thus the whole rig was set in motion; by winding up the rope the tightener was raised, the belt slackened, and the pulley with all its dependencies would remain at rest. It is claimed in favor of this mode of belt gearing that it relieves the spindle of the side strain caused by the tightening of the belt; this strain, however, is not got rid of, but only transferred from the upright spindle to the horizontal shaft, which is more liable to cause trouble than the spindle itself. These horizontal bearings are pretty large, and they revolve at such a velocity that they require constant care and attention to keep them oiled, the oil being liable to leak off; we have seen a set of the brass boxes in which they run, cut out and destroyed in a very short time, by an occurrence of this kind, the brass being thrown out in particles like filings and resembling gold dust.

To reduce the velocity of these shafts and obviate this tendency, Mr. Smith made the bevel wheel on the shaft larger by one-third than the one on the spindle, but this, while it reduced the velocity, increased the pressure in the same proportion, and in a measure neutralized the intended advantage. It is found in practice, that a spindle and millstone driven by a pulley and belt instead of a cog-pinion, is not disturbed by the side strain of the belt, if the pulley be large enough; and it should never be less than thirty-three inches for an ordinary sized stone, and as much larger as may be desired.

A belt will run on a large pulley with a swift light motion, perfectly cool and safe and perform a certain amount of work with ease and satisfaction, and without any perceptible injury to the belt; but reduce the size of the pulley (and drum of course to match) say to one-half, and try to do the same work with the same belt. It must be very tight, still it will slip, and heat, and burn, and tear out the lacing and rivets, and run off the pulley; in short, it will be a continual bother as long as it lasts, and will soon go to destruction. The extra strain upon the bearings is not the least of many disadvantages. With a large enough pulley on the spindle, the side strain of the belt is only sufficient to keep the bearings of the spindle at the toe and collars always firm to the same side, and insure a steady smooth motion, when these bearings wear loose and would otherwise have some side play; this is the reason given by millers and mill owners who have tried both cog gearing and belts, for preferring the latter. The cog gearing is generally the cheapest, it is much less troublesome to keep in order, and more durable, but the belt motion is so much smoother that they get from one pound to one and a half more flour from a bushel of wheat, and the dress in the stones will last considerably longer.

We once geared an overshot mill with the water-wheel and drum shaft exactly like those of Mr. Smith's, but we dispensed with the horizontal shaft and bevel gearing. We placed a pulley on each spindle, and fixed the upper surface of the drum exactly on a level with the centre of the sheave on the spindle. The driving belt passed straight and level from the driving side of this pulley to the top of the drum, taking a quarter twist in that distance, *i. e.*, it left the pulley upon edge, but fed on to the top of the drum flat; it passed around the drum and up over a tightener set at an angle which returned the quarter twist to the belt, and fed it on at the proper height and up on edge ready to pursue the course indicated.

It requires some calculation and care to locate these different parts aright with respect to each other, but when this is attained, the belt works just as well this way as any other; and we never geared a mill before nor since, where the same expense of money and water would do more work, if as much. It may assist any one intending to run a belt in this way to note, that a plumb line hung at the driving side of the pulley should range exactly with the centre of the drum, and a level set on top of the drum should point to the middle of the sheave; this allows the driving belt to work straight and without angle or slip, and leaves all the angle to the slack belt, where it is accommodated by the incline of the tightener which directs it in the proper position and height to feed around the pulley. This plan of belt gearing is, in common with Smith's, subject to the objection that the weight of the slack belt hanging upon the revolving drum causes a useless wear upon the belt and drum, injurious to both, and a slight waste of power. To obviate this, we have fixed a convenient plan for hitching the belt up clear of the drum when slack, which answers well.

Another overshot mill of four run of stones is geared by Mr. Messenger, a contemporary of Mr. Smith, and from the same "school," in the following manner: The wheel is thirty feet in diameter, and set down in a natural gulley in the rock, improved by blasting, almost to the shaft; a cog gear is put around the outside of the rim next to the interior, and the motion taken from this by a heavy crown wheel or pinion placed in the centre at the descending side. The shaft of this pinion is extended the whole length of the mill, just above the lower floor, on the timbers of which the shaft bearings are fixed; this shaft from the pinion to the next bearing is about eight inches in diameter; here a bevel wheel about three feet in diameter is placed upon it, driving a pinion of about two feet, on an upright shaft. From this bevel gearing to the next bearing the lying shaft is about six inches in diameter, and here another set of bevel gearing and an upright shaft are placed, like the first; at this point another portion of shaft about four inches in diameter is coupled on, extending about twelve feet further, where it drives another and smaller set of gearing and upright shaft which carries the smut machine and cleaning apparatus. The other two upright shafts have each a double drum, from which two belts with their bearers and tighteners run off in opposite directions to their respective stones. These shafts, running as they do along the centre of the mill, admit of an equal and convenient distribution of the stones and bolts, a bolt being placed in each corner of the building, and the stones so located with respect to these as to communicate by the elevators in the most direct manner.

The mill floor is just above the wheel and spout at that end, and a little above the mill-yard at the other; the lower part being of stone and the upper of frame. This is a good mill, but upon a very small stream, and is often deficient in water, and this defect is somewhat aggravated by over gearing, from the old but mistaken notion that the slower an overshot wheel turns the more work it will perform with the same water. (See a previous chapter on water-wheels on this subject.) The stones make about ninety-six turns for one of the wheel, and to accomplish this the pulley on the spindle is made too small. To compensate for this, the belts are made exceedingly wide and heavy (from nineteen to twenty-four inches wide and very thick). Still they would slip and heat, so that the owner told us that he insisted upon the pulleys being made six square to enable the belt to hold upon them; this, however, the millwright refused to do, and although the owner assured us that he would still have it done, we think the foolish experiment was never tried. The extra stiffness and weight of a belt beyond what is necessary involves a slight additional waste of power. It requires a certain force to bend a belt short and suddenly, and a like force to straighten it again or bend it in the opposite direction, and this force is in proportion to the weight and stiffness of the belt. Now there are in each belt three points where the bending and three other points where the straightening process is constantly progressing like the undulations of a wave, and with a rapidity equal to the velocity of the drum, and sheave, and tightener, as the belt enters upon and leaves each of these. This waste, with an ordinary belt and ample motive power, is not taken into account; but when the power is limited, and the belt very stiff and heavy, it will be found to detract very sensibly from the working of the machine.

We once astonished a friend who was incredulous and "couldn't see the point," by stretching a large heavy rope tight around several pulleys, turning freely on their bearings, to one of which a hand-crank was attached; we requested him to turn the crank until a velocity was imparted to the rope equal to that of the belt in question. The experiment satisfied him that a great power was absorbed somewhere, but he was still incredulous as to the actual point of absorption.

The last illustration we will give of the various ways in which overshot mills are geared, is a small mill of two run of stones, all the movements, heavy and light, being made with cast-iron cog gearing, to the total exclusion of belts. This mill was geared and is owned by an old Scotch millwright. The wheel is twenty feet in diameter, and is partly sunk under the foundation like the last one, by blasting out the rock; the wheel inside the mill, and with the spout, comes up to the mill floor; an extra rim is bolted on to the arms of the water-wheel fourteen feet in diameter and about six inches thick; a spur gear is put around the outside of this rim, and a pinion gears into it at the descending side at the centre. (Both this wheel and that last described are placed down stream sufficiently to bring the shafts to the centre of the building, and turn up stream towards the entering water, and technically called "pitchback.") This pinion drives a horizontal shaft, on the other end of which is a bevel wheel working into another the same size (mitre wheels) upon an upright shaft. On this vertical shaft is a cast-iron spur wheel six feet in diameter, working in a pinion upon each spindle about sixteen inches in diameter, which drives the stones. A small shaft is coupled on to the top of the spur wheel shaft and is carried up to the next floor; this shaft, by a series of small wheels and shafting which intervenes, carries both bolts, with their respective elevators, and the smut mill grain elevators, together with a variety of saws, lathes, and other machinery in an adjoining workshop; near the lower end of the spur wheel shaft is another set of bevel gear, driving a long horizontal shaft which extends to another workshop, at the other end of the mill, containing larger circular saws, and a planing machine; from this last shaft the smut machine is driven.

This is a substantial little mill, the movements being wholly of iron; but, like the one last described, it is overgeared. The circumference of the water wheel moves only three feet per second, and acting upon the pinion with a lever purchase as it does, the pressure upon the cogs and bearing of this small pinion is so great that they are compelled to keep the gearing constantly greased, and being so nearly in contact with the water, it is a constant source of trouble and expense. With all this care, the pinion is soon worn out and has to be replaced by a new one. We tried to persuade the owner to enlarge this pinion, and thus increase the purchase, and allow the wheel to move with a lighter, freer motion. But this is the millwright referred to in a former chapter, who wants "thirty yards of the stream on his wheel at once." We have since been informed that he enlarged the pinions once or twice when he renewed them; this would indicate that experience has taught him the same lesson that we failed to impress upon him. This mill, like most of those described, is built of stone to the main floor, which is above the level of the mill yard three feet, the upper part being a wooden frame building one story and a half high.

From a review of all these various methods of gearing an overshot grist-mill, not to mention the still more complicated methods now nearly obsolete, of extending the water-wheel shaft and gearing from it by a pit-wheel, &c. &c., it is obvious that the simplest possible arrangement involves such an amount of cumbrous and complicated machinery, expensive at first, and requiring continual care and expense to keep the whole in working order, and replace the wear and tear, that the overshot wheel should not be used where the water power is sufficient to do the required amount of work with any of the simpler wheels. It may be further observed that great additional expense and inconvenience are incurred in making room for this kind of machinery, either within the mill itself, or in an additional wheel-house, and in conveniently adapting the whole to the situation—the mill-yard and other approaches; whereas all the varieties of modern wheels adapted as they are to run upon vertical shafts, admit of so much uniformity in their application to grist-mills, that the building may be planned for all other conveniences without regard to the particular kind of wheel to be used, and any of these that may be selected as the most suitable in other respects, will answer the building; a longer or shorter shaft furnishing the necessary accommodation. Or one kind of wheel may be taken out and replaced by another without materially disturbing any of the other arrangments.

We began this particular division of the work by trying to give a description of the building necessary, or suitable for a grist-mill, but have inadvertently wandered so far among the mazes of these wheels and the dark recesses containing them and their machinery, that we have lost all hold of the "thread of the discourse," and find it impossible to get hold of it again; we must therefore dismiss the subject with a few general observations applicable to any custom-mill building, and proceed to give some account of the various parts required to constitute all the movements and fixtures inside of an ordinary grist-mill. We have already observed that the lower story of the building up to the floor containing the stones, should be of stone if practicable; if not it should be as solid and substantial as possible.

This part, unless for some special reason already referred to, need not be more than eight to ten feet high; all the husk timbers supporting the stones and machinery, as well as those supporting the floors in this story, should be placed so that they may be taken out and replaced by new ones when they decay, without disturbing the walls or upper portion of the building. Second, the building above this floor containing the stones, bolts, &c., being away from the water, and only required for storage and shelter, need not be so substantial, and may be of any material that taste or convenience may dictate. This story should not be less than nine or ten feet between floors, to allow sufficient height for bolt chests, hoisting crane, &c. The chamber, if not intended for conveyor or bolts or something extra, need only be high enough for convenient grain bins or other storage. If the whole building be of stone, or solid brick walls, the timbers should all be placed so that in case of fire they might burn and fall off without disturbing or prying down the walls. We owned a grist-mill built of sandstone, that was burned a few years ago. The walls were but little injured. The wall up to the main floor was thirty inches thick; here a ledge of six inches was left, and the timbers rested upon this ledge at each end and a heavy trimmer supported on a stone pillar at the middle. The wall through this story was two feet thick, and again six inches were left to support the next timbers, above which it was only eighteen inches; a stone lintel was placed over each door and window outside, but extended into the inside of the wall with wood; the whole interior of the walls was plastered with a heavy coat of mortar upon the stone without lathing. The fire cleaned the woodwork out to the foundation, but did little damage to the walls, except that it peeled off the plaster (which had saved the stones from injury) and damaged the inside of the door and window heads; six days' work of a mason renewed the walls and chimney tops nearly as good as ever.

This mill was driven by an overshot wheel twenty-seven and a half feet in diameter, placed in a separate wheel-house, built of stone, between the mill and a bank thirty feet high, upon which the water was brought by a canal. The wheel was kept running thoughout the whole continuance of the fire, and was saved, although the upper story of the mill was occupied by a Daniel's planer and several circular saws and lathes, and also contained the whole woodwork for a new building, with the exception of the frame; the burning timbers fell down upon it and around it, at times stopping it altogether, but it would gain in power as the buckets filled and started again. Timbers supporting the flume finally gave way, and the whole remaining fabric fell upon the wheel and was carried down with a crash, sending up a volume of fire, steam, and cinders like a volcano. A temporary board spout was applied to send on the water, and the wheel when all was over was none the worse except one broken arm, and is now as good as ever. This fire occurred in October, and although we had no insurance, and but little means for such an emergency, by saving the walls and the wheel, we were enabled by the assistance of the neighbors to have the mill in operation again by the first of January.

We mention these facts in order that others may profit by our experience, in the construction of new mills, as we have seen two stone grist-mills that were burned since, where the walls were almost totally demolished by the timbers being built into them. When these timbers burned so that they fell, their remaining weight acting with a lever purchase by their ends in the walls, brought the walls down with them, and nothing remained but a pile of ruins. The walls built in this way are so much thicker at the bottom than at the top, that they should be built with a regular incline outside, especially where one *side* or end is built against a bank, to balance the incline or "batter" as it is called inside, and bring the centre of gravity upon the foundation equally. This precaution may be disregarded in an ordinary building, and the outside carried up plumb, but in a mill it is essential, as the jar and tremor caused by the machinery affect the building, and give it a tendency to crack, and especially to bulge out. And, third, as to the size of the building: this-last mentioned mill is forty-five feet long by thirty-five feet wide outside, and is large enough for three run of stones, with their bolts and other appendages. We have geared some small mills, with only two run of stones each; two of these are twenty-six by thirty-six feet, and two others are thirty-four feet square. One of these last has had a third run for grinding feed put in since, one of the bolts being removed upstairs to make room, and provided with a screw conveyor and spouts to bring the flour down again. This mill does a large amount of work, both custom and manufacturing, but lacks room. It is to be enlarged the ensuing season. Several mills with three and four run of stones around here, are forty by fifty feet; others with four to six run, are forty by sixty feet. Much depends upon the convenient grouping of all the different parts; a proper attention to this makes some small mills appear roomy and commodious, while its neglect or a confused arrangement of the machinery and fixtures in others of a large size, makes them appear crowded and unhandy.

The Husk Timbers, &c.

These consist of two parallel sticks of large timber extending from side to side of the building, and three or four feet apart, placed at the level of the floor, immediately under that upon which the stones are placed, and directly under the stones. These foundation sills should be solidly supported, either upon stonework or sufficient posts, and the bridge tree posts tenoned into these; these posts extend from these sills up into and support the meal beams, which are two very large sticks running parallel with the first two and placed under the next floor. Upon these meal beams the bed stones are placed, and exactly under each bed stone are the two, or sometimes four bridge tree posts already alluded to, which support the bridge trees upon which the spindle stands. The bridge tree is hung upon a pivot at the back end, and the front end is supported by a lighter rod; a rod of iron attached to a tenon on the end of the bridge tree, and running up through the floor to a convenient height where it terminates in a screw and nut. On this nut a small hand wheel or other convenient wrench is placed, and by turning this in one direction, the lighter rod and bridge tree, with the spindle and runner stone are let down, and by turning it in the opposite direction they are raised up.

Fig. 37.

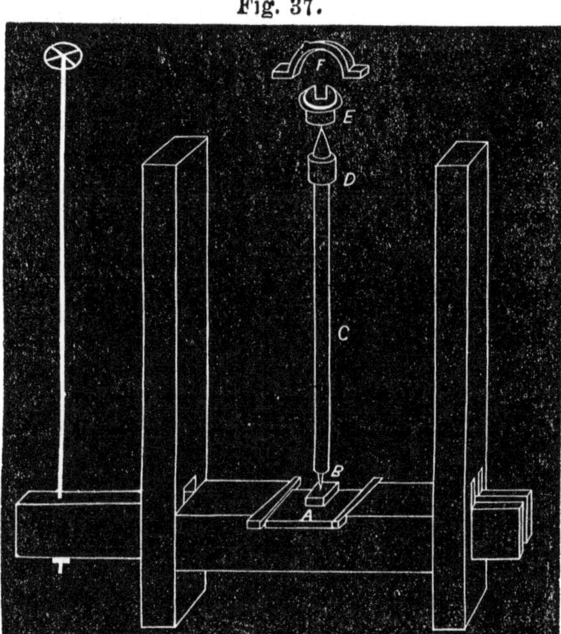

A step; *B* toe; *C* spindle; *D* collar; *E* driver; *F* bail.

When the bridge tree is supported by a single post at each end as in fig. 37, then the back end is made with two tenons, two or three inches thick, according to the size of the stick, and these are put through corresponding mortises in the post; a counter mortise is put through the post in the opposite direction, about two inches wide and five deep at the lower end of the other two. This cross mortise is made square at the lower end, but round at the top, and a key (or two heads and points are better) is made to fit this mortise in thickness, but slack in width, and tapering. A half-circle is cut out of the lower edge of the double tenons on the bridge tree to fit the round top of this key; the tapering key is driven into its place under these tenons, until the bridge tree is raised up and rests upon the round edge of the key which keeps it in place, at the same time forming the pivot for this end to work upon; the front end has a single large tenon let through a suitable mortise in the front post, in which it is raised or lowered at will by the lighter rod. These mortises in the fender posts must all be deeper than the depth of the tenons, to admit of play, and the bridge tree must be kept as nearly level as possible by driving or slacking the key under the back end.

When four bridge tree posts are used, the two behind are placed one on each side of the bridge tree, and a short bar is mortised across between the two close above the bridge tree, the end of which is hung up to this bar by two sets of screw eye-bolts; by slacking or tightening the nuts on these eye-bolts, the bridge tree is kept level. The front end passes between the two front posts, instead of through a mortise in the single one, and is guided by the lighter rod as in the other case; the two posts are connected by screw bolts, which are tightened if the bridge tree gets loose by it and the posts shrinking by seasoning.

Although it is essential that the bridge tree should

have a free chance to raise and lower, it is equally important that it should be perfectly solid every other way. A board may be placed between each pair of these posts in a groove made near the outside corner of each for that purpose, one board passing behind the back end of the bridge tree, and the other having a mortise slit for the lighter tenon of the bridge tree to pass through; this will both help to steady the whole concern when the bolts are tightened, and improve the appearance.

In the upper side of the bridge tree, and near the middle, a piece is cut out about two inches deep and a foot wide, into which the step for the spindle foot is placed, and fastened by four dovetail tapering keys; two of these go across the bridge tree in the notch for the step, one on each side, the other two lengthwise of the bridge tree, are driven between it and a projecting ledge along each end of the step which runs over each side of the bridge for that purpose. By slacking one of these keys and driving the one opposite, the step is shifted in the direction necessary to "tram" the spindle. This step has a square or round box on the centre, called the oil pot, about four inches wide and the same in length. A flat piece of steel is placed in the bottom of this, and another similar plate with a hole through the centre is placed above the first, with a space between for oil, and also some provisions made for the free passage of the oil through the upper plate. The cast-steel toe or pivot upon which the spindle runs passes through the upper plate which keeps it in place, and turns upon the lower one; both these and the pivot being hardened to the same temper. The upper end of the spindle where it passes through the eye of the bed stone, has an enlarged collar about seven inches long and turned true, around which the boxes are placed which secure the spindle in the eye of the bed stone, and furnish its upper bearing.

Mill spindles were formerly made of wrought iron with this collar of steel welded on; the steel toe was also welded on, and they were more expensive than the cast-iron spindles now used, and did not keep so true. The welded collar had always hard and soft portions, and soon wore unequally, while the cast collar wears equally all around and keeps true. The welded toe is also inferior to the modern practice of turning a toe of cast steel and setting it in a tapering hole drilled in the end of the spindle. These require to be well fitted to bear equally the whole distance; the point must not touch the bottom of the hole, as in that case it will get loose, and spoil the spindle in a short time. It is customary to wash with pickle or other stuff to cause the toe to rust in. An inexperienced mechanic will generally get the point or toe too large for the guide plate in the step; this is a cause of frequent annoyance and mischief, and occurs by rimming out the hole in the plate to fit the turned pivot when both are annealed and in a soft state. Then in the process of hardening, the shrinkage of the large plate diminishes the size of the hole more than the small pivot is diminished, and it will be found too tight or large for the hole, unless some allowance has been made; but this is not all, for the hardened pivot may be ground down until it will turn in the plate, and when first set in motion, with the weight of the mill-stone, spindle, &c. upon it, and the friction upon the bearing plate, the pivot will heat and expand and pinch in the guide plate, which increases the heat and expansion still more, while the plate having no weight or friction upon it, except at the pivot, is not heated or expanded except at that spot; but there the heat will increase so rapidly that we have frequently seen the pivot welded into the step within twenty minutes after starting. When this occurs it must either turn on the pivot inside of the spindle or shaft, in which case the hole is soon spoiled, or it must break the pivot, or stop the machinery; we have seen all these issues from that occurrence, and had our own share of trouble with each.

In one instance in starting a new overshot mill, the pivot of a spur-wheel shaft driving three run of stones welded in this way in this guide plate, and turned in the shaft. The crown-wheel or pinion driving this was only twenty-six inches in diameter, and therefore the side strain upon the pivot was considerable; it had only run about half an hour altogether when it was noticed, but in that short time the pivot had worn for itself a good half inch of play. To remedy this we had to raise up the shaft and take out the pivot and step together, then heat the guide plate and pivot to a red heat, and cool the pivot suddenly to shrink it while the plate remained hot. By this means we got them apart but with small fragments of each splintered off and sticking in the other; we then took a piece of cedar that was soft and fitted it into the enlarged hole in the shaft by filing away the marks and rings shown upon this, by turning it in the hole, which, although ridged, was still tapering, we got an exact pattern of the hole. By this pattern we had a new pivot turned, and perfected the fit by grinding it into the hole with emery and oil; this was done by putting a wooden pulley on the lower end of the pivot to give it motion, and pressing it tightly up into the shaft, adding more emery during the operation until the fit was complete. This made a good job, and although it was considerable trouble and expense, it was much less of either, than to demolish the whole concern, and send the shaft back to the machine shop.

In this connection we will mention another trouble we had with a similar spur-wheel shaft. This was an old shaft and pivot, the step and guide plate only being renewed; the guide plate, as in the other instance, had been fitted too tight to the pivot, and they heated and welded, or stuck fast. In this case the shaft and pivot had run so long that the heating and cooling so often repeated, and with so much jarring and weight upon them, had settled the two together so firmly that the pivot could not turn in the shaft, but broke off outside, and the problem was how to get the piece out, to put in a new pivot. We tried the usual recourse in such a case, *i. e.*, heating the shaft, but without effect, and had to drill a small hole into the shaft at the upper end of the pivot, and prick in gunpowder through this hole until the cavity at the end was filled; the explosion of the powder removed the obstacle. When anything of this kind occurs with a spindle it is easier rectified, as it can be taken out more readily, and sent to a shop for the necessary repair.

The spindle head extends about six inches above the collar and bed-stone, this part tapering, and four, six, or eight square, has the driver which turns the running stone fitted on to it, and a small steel pivot drilled into the top, on which the bale and running stone are balanced. This pivot, or cock head, as it is called, is rounded off to a point, and a corresponding depression is countersunk in the bale to fit it. This cock head should not be tempered, as it has sometimes to be turned anew, or altered in shape to fit and balance the bale upon it; this can be done by fixing a solid rest of any kind at the point, and applying a steel cutter, or if but little is to be taken off, a sharp file held against the point by this rest will turn and fit it. We have drawn the temper of these cock-heads, to refit the bale, by heating an old driver, and placing it upon the square spindle head, until the latter was well heated, and then applying the hot bale to the pivot, for some machinists persist in hardening the pivot although there is no friction upon it, except balancing the weight of the stone, and when the rest of the spindle is all right, there is no other necessity for taking it out. We have frequently drilled the spindle head for the pivot by keying a suitable drill into a piece of plank or other lever, and shoring this properly down with the drill to the centre of the spindle, which is then set in motion, until the hole is of the proper depth and shape.

The bed stones are placed upon the meal beams, the large upper timbers of the husk. Some merely notch a couple of pieces of short plank or timber across between these beams at a proper distance apart, and lay the back of the stone flat upon them, inserting thin wedges between the back of the stone and these pieces and beams, to raise and level the stone. A better way is to make a square frame of four wide pieces of plank, framed together at the corners, and fit this frame to the back of the stone, either by planing off the high spots, or by interposing a thin bed of soft mortar between. Under each corner of this frame cut a space out of the meal beams to admit a wide key, and by driving or slacking these keys the bed-stone is levelled and kept true. We have seen the bed-stones hung upon the point of four large screws, one at each corner of such a frame, but prefer the keys. The stone should be kept free and clear of the floor, and the space between covered by a rim of plank about six inches wide, fitted flour tight all around the stone, and resting upon the floor, the outer edge being champered off to diminish the jog. The curb rests upon this rim, and three or four short pieces of circle, wide enough to fill the space between the stone and curb are nailed on this rim, at equal distances and close to the bed-stone to locate the curb and keep it in place. The precaution of keeping the bed-stone clear of the floor, and dependent wholly on the husk timbers, is necessary to prevent the working of the stones from being disturbed by the yielding of the floor, when large piles of bags and grain are placed upon it, and raising when these are removed. The spindle and running stone rest upon the bridge tree, and the space between it and the bed-stone while grinding is so minute and so accurately graduated that a rise or fall of the bed-stone by the floor, of only a small part of a hair's breadth, would make the flour too coarse or too fine.

The running stone has a bale sunk in from the face of the stone by cutting two grooves in opposite sides of the eye, and a little larger than the ends of the bale, to admit of moving and making it true to the centre of the stone. This is done by a sweep stick half the diameter of the stone, with a piece attached to one end, reaching down the eye of the stone, to the countersunk centre of the bale in which it is fitted. By keeping the point of this steady in the bale, and moving the other end of the sweep around the skirt of the stone, it will show which way to move the bale to bring it to the centre. When fairly in the centre, and true with the face of the stone, it should be fastened by wedges to keep it steady, and the bottom and skirts of the grooves closed with clay or mortar, and the spaces run full of lead. Great care should be taken to lay out these recesses for the bale true to the centre, and of the right depth. The first is done by fitting a board in the eye and making the centre on it by a sweep and laying out the recesses or grooves by that centre. The same centre will be needed to lay out the grooves for the driver, if that kind is used, and also to lay out the dress in the stones. The proper depth for the bale might be found either by hanging it in its place on the pivot of the spindle, and measuring from the ends to the face of the bed-stones, which will give the depth to which these should be sunk below the face of the runner, or by laying a straight edge upon the pivot of the spindle, and measuring from it down to the bed-stone: which will give the depth from a straight edge laid upon the face of the running stone to the socket for the pivot.

There are several ways of fixing the driver. A very common way is to place a straight driver of either wrought or cast iron upon the tapering part of the spindle head; if that be square, the squares secure it; if round, it is secured by a key. This driver is placed at right angles with the bale, under the bow of which it passes, and is let into a recess in each side of the eye, cut for that purpose. The ends of the driver, however, must not be fastened in the stone like those of the bale, but must have considerable clearance to allow the stone to balance freely on the bale. Particular care is necessary to get both ends of the driver to catch the stone exactly alike, because if one end touches a hair's breadth before the other, it will tip the stone up on that side, and bring the opposite side in contact with the bed-stone, and a constant vibration and rumbling are the consequence, which almost every miller has observed, and in many instances is wholly unable to account for.

The ordinary method of locating this driver is as follows: Place the runner stone face up and level; then secure the driver firmly in its place on the spindle, and stand the spindle up plumb, with the lower pivot up and the upper pivot down in its socket in the bale; make it secure in this position, with the ends of the driver in their respective gains in the stone. Next fix a tram, one end with a bearing on the toe of the spindle, now turned up, the other end to sweep the face of the stone close to the skirt. Fix another piece, one end with a half-circle to fit and bear upon the collar; the other end fastened to the end of the first piece, at the skirt of the stone. By turning this tram round, and shifting the toe and spindle until the tram sweeps the face of the stone all around equally, the spindle is gotten true with the stone, and both are in the same position with respect to each other which they will occupy while grinding.

You may now place the iron boxes or "thimbles" which line the recesses in the stone so that they will both touch the ends of the driver exactly alike on the *driving* side, then move the ends of the driver back, and fit them by chipping, filing, or otherwise, to the back side of the thimbles also. This back fit is only of use when the "backlash" caused by stopping the machinery occurs; in this case, the stone being raised clear of the bed-stone and no grain between, becomes the driving power, until its accumulated momentum is overcome by the resistance of the machinery. And while this is going on, if one end of the driver only touched the stone, it would tip on the bail as in the other case, and spoil the dress and face of the stones. With all this care, in fitting the bearings of this kind of driver, it will in some instances still be imperfect, and may be completed by marking the inside of the thimbles with paint, after the spindles and other things are in place, and letting the stone carefully down upon its place without touching the driver; then turning it each way until the paint marks the points of contact upon the driver, and then taking up the stone carefully, and filing off the paint marks. Of course the thimbles are fastened securely in the stone before this operation is performed; these are fastened like the bale by securing first with iron wedges, and then running the cavities full of lead.

We once fixed a driver of this kind that had baffled several first-class millers and millwrights, and rendered the best wheat stones in the mill useless for several years. The stone had been rehung and balanced anew during this time by several adepts, and, among others, by the celebrated Mr. Brown, mentioned elsewhere in this work, as having reduced the balancing of mill-stones to an invariable rule, which before had to be accomplished by repeated trials. The real source of trouble had escaped their investigations, and by some it was attributed to witchcraft. We examined the machinery, and everything appertaining to these stones, and found all in perfect order, except that the driver was worn upon one end more than on the other. This end was chipped and filed, and by a few applications of paint as above directed, and a little more filing, we succeeded in making a perfect fit, which may be known by both ends equally holding a slip of thin paper inserted between these and the thimbles. We started the stone, it being just clear of the bed-stone, and saw that it ran true, and was in perfect balance both running and at rest, and told the owner he might now put in a grist of wheat and we would warrant it to grind all right. He advised letting the stones come together before the curb was put down, as he had seen them run just as well before, when high and free, and the witch only interfered when the stones were set to work. We took his advice, but whenever the runner touched the bed-stone it began to rumble and shake so violently that it stopped the stone, water-wheel, and all in a few revolutions, and with a noise like thunder. We then made a close investigation of the driver and spindle, which were round, and the driver fastened by a steel key next to one end. That end was held tight by the key, and the whole driver was apparently tight, but by striking the other end with a heavy maul, first on one side, then on the other, we discovered that the hole was too large for the spindle, and allowed that end to shift when the strain of grinding came upon it. This was the cause of all the trouble, and the owner remembered that the original driver of that run had been broken and a new one cast for it just about the time it took the caper; we cut another key seat, and put another key in that end, and fitted the driver anew, and the trouble ceased.

This kind of driver is in general use throughout the country, although in addition to the practical inconveniences already enumerated, there is still another, which is, that the ends of this driver, entering the stone at right angles with the bale, causes four obstructions to the grain entering the eye, instead of two only, by the driver next described. This makes the grain, and especially feed or foul grain, more liable to lodge and be carried round by the centrifugal force, and thus clog the eye, which is a source of considerable annoyance. The next driver to be considered, and the one which we prefer, consists of a round block of cast iron, fitted on to the tapering top of the spindle, which for this purpose should be eight square, and reaching from the collar nearly to the upper surface of the bale. The top of this block is round, something like the round crown of a "Jim Crow hat," and has a recess cut out across the top to admit the bale, which should hang free in this recess, touching the pivot upon which it balances. This bale should be made deeper and thinner than the ordinary square one, which allows more strength and room to the driver. This can be fitted easily and perfectly, both on the driving and backlash sides, before the bale is fastened in the stone, and being in the centre of the eye, in sight, and within reach, there is no trouble in getting the bearing alike on both sides, and when once fitted, there is no further trouble. This plan, we need hardly say, drives by the bale, and leaves the eye with only two obstructions.

In some instances where the eye of the stone was very small, we have had the driver very small, and cast solid with the spindle, or a wrought iron band has been shrunk on to insure sufficient strength. This we would not recommend, as the fixed driver is in the way of placing or removing the bush, boxes, and collar, around the spindle, in the eye of the bed-stone. It is better to have it movable, but as small as is compatible with sufficient strength; the strength may be increased by banding with wrought iron, as indicated. There are several other forms of drivers used which need not be described. One of these is similar to that first described, but shorter, and instead of being let into recesses, cut in the stone for a driving hold, it clutches upon the lower corners of the bale. When stopped, this allows the stone to turn nearly a half revolution before it backlashes, and when started the driver comes around the same distance before it comes in contact, which causes a shock in starting the stone which is injurious to the machinery.

Another kind, a modification of the last, has a clutch on both sides of the angle of the bale; this is an improvement, as it is easily fitted, both for driving and backlashing, and obviates the mischievous shock induced by the other; but it obstructs the eye more, and the clutch end in front of the bale forms an angle with it, in which the grain lodges and forms a nucleus for clogging, which is encouraged by the centrifugal force, and is sometimes a source of annoyance.

The "Stiff Ryne" is not used to any extent, at least in this country, except for hanging the stones used for shelling and grinding oats in the manufacture of oatmeal, and will be described under that head. The next machine in order is the Dansil. Its business is to shake the shoe, and induce a regular feed of the grain, and for this reason it is sometimes called the feeder. It is coupled on to the centre of the bale, and stands directly above and in line with the spindle, the upper end turning in a bearing made in the centre of the cross-bar of the frame supporting the hopper. It is generally attached to the bale by a crotch, or forked end, which straddles the bale in a recess made to fit it; and to insure accuracy a small hole is drilled in the centre of the lower end between the forks, and a corresponding hole is drilled in the centre of the bale. A short dowel-pin of iron is placed or fastened in the bale, and the hole in the end of the dansil dropped on to the projecting end of this dowel secures it in place. Sometimes the dowel is fastened in the dansil, and dropped into the hole in the bale; but this should never be done, as the hole in the bale is liable to get filled with dust and grain, and is troublesome to clean out, while no occurrence of that kind takes place with the dansil. There are several ways of making the body of the dansil. An old-fashioned plan is to weld three, and sometimes four suitable rods together, leaving six or eight inches at the height of the shoe not welded. These are spread apart equally, and as they revolve in contact with the end of the shoe, they increase the shake and feed of the grain. Another plan is to place small iron disks at a suitable distance apart; these have six small holes drilled near the circumference of each, and at equal distances apart, and small polished rods are inserted in these holes reaching from one disk to the other. These have a hole in the centre, and are slipped on to the dansil and keyed at the proper height; the dansil in this case being a round turned rod.

This rig resembles an old-fashioned spindle, with its trindle or lantern for the wooden cogs, on a small scale, and being all turned and polished presents a fine appearance. But it is more fanciful than useful; things sometimes get caught in these little rungs; in one instance the button of a miller's shirt sleeve was caught, and the sleeve taken off to the shoulder in an instant. Had the material been strong enough, it might have taken the arm with it. Sometimes the grain strikes in among the lower ends of these rods and is thrown over the stone: we know several mills where they have sewed a covering of leather over these rods to guard against such occurrences. Another plan is to fasten four small rounds close upon a round central shaft like the last one; this is done by drilling holes through the shaft at suitable distances, and turning the ends of these rounds at right angles, and inserting them in these holes where they are secured by riveting. This arrangement resembles the old-fashioned wooden dansil with iron knockers now nearly obsolete, or only used in some old oatmeal mills. The cheapest, and probably the best dansil, is simply a piece of iron about an inch square, with the ends as described, and the sharp corners slightly taken off, and the whole finished in a workmanlike manner.

The bush in the bed-stone through which the spindle passes is frequently made of a skeleton framework of iron, with suitable spaces for the four wooden boxes and keys which form the upper bearing of the spindle. The form of this skeleton bush is so nearly like the wooden one, that a description of each would be needless, and we will only describe the ordinary wooden bush. To make this, take a square piece of wood large enough to fill the eye, which is generally about ten inches, and cut it off a little longer than the thickness of the stone. Fit this into the eye, then take it out and centre both ends, and from this centre strike a circle mark, a little larger in diameter than the collar of the spindle; strike another circle upon the upper end the exact size of the spindle, draw two parallel lines square across the end, each cutting the outsides of this last circle, which will make them just the width of the spindle apart; now draw two other lines exactly like these across in the other direction, or at right angles with them, and extend all these lines down the outside of the block four inches, and square these across on all the four sides to this depth. The centre must be cut out to the largest circles to admit the spindle, and the four spaces cut out to these last marks, to admit the boxes and keys.

To make the boxes, take a piece of poplar or bass-wood, or any soft wood, and dress it to the size of the collar one way, and to the depth of the recesses, or spaces which we have made four inches in this case. Set the compasses to the size of the collar and mark out the four boxes on the top of this piece, one end of each being circled by the dividers, and the other square, and allowing half an inch short at the end of each for keys; now square the stick across, and lay them out all the same on the opposite side and work them out. The bush having been slipped over the head of the spindle and driven down into the eye to its place, these boxes may be put in, and two wedge keys at the end of each, one with the head down, and the other with the point down. These last four may now be driven lightly and equally down until all is tight; care being taken to have these wedges the right size, so that they may go to the bottom of the boxes without being too tight, which would cause the spindle to heat and burn loose. We forgot to mention that when the boxes are to be tightened by these wooden wedges, a space should be cut out of the bush, from the box down through, for the last key at least, and for convenience in driving them out. We rather prefer to leave the bush whole below the boxes, and instead of wooden keys pack the space between the end of the boxes and stone with something more elastic than wood, which is apt to get too tight at first, and cause the spindle to heat and swell. When this occurs, the spindle will be loose when stopped and cooled, and will continue to get worse, unless tightened again immediately. With elastic packing, the spindle will not heat so as to make it swell and shrink, and if it did, the packing would close the boxes around it again, and save the trouble of taking up the stone to tighten it.

Several different substances are used for this packing, India-rubber being the best. Old rubber boots or shoes answer well; these are cut in strips as wide as the boxes, and driven in with a wooden rammer. Strips of flannel or any woollen fabric may also be used. We have used woollen flyings from the carding-mill, and several packed in the following manner many years ago remain as good as ever, only requiring occasionally to be rammed down and a piece added to fill up the space. We cut an old linen salt sack into strips the width of the boxes, and began by pushing one end of the strip to the bottom of the space, and continued pushing down six or eight inches at a time, and ramming till the spaces were packed full, and then fastened a little piece of wood on top of each by a small nail. When the boxes require to be taken out, draw the nail and remove the piece; then take hold of the end of the strip and pull it out, if more remains take a sharp hooked point of some kind and hook the next end until all is pulled out. It is handier removing these when each space is filled by a single piece, and if one piece is not sufficient they may be sewed together; when fed in and packed regularly, they come out quite freely, and more time is often spent in getting out a single wooden wedge than all of these.

There is a difference of opinion among millers as to the proper height to place the bush. Some want it higher than the face of the bed-stone, and the corners rounded off to clear the ryne and running stone; others want it just level, and are very particular to keep it just so, while another class must have it considerably below the level. The first is defended on the plea that it leaves no chance for the lodgement of the grain, which is swept off clear by the irons at every revolution, and for this reason is preferable where small grists of different kinds of grain have to be ground in the same stones. The last has a recommendation exactly opposite and equally important, which is that the space below the level of the face of the stones, and below the reach of the irons, by being filled with shorts or other soft stuff, will catch nails or thrashing-mill teeth and nuts or other iron that get accidentally into the grain, and thus spare the damage and danger occasioned by these getting between the stones. These sometimes break the stones and cause much mischief. The skeleton metallic bush mentioned, will be understood by its figure, and the boxes can be fitted in by any miller who has the use of tools; a vacant space in each corner has also to be filled with wood to nail down the leather cover or collar. Some of these have patent metallic covers, as well as metallic boxes of various kinds; wood, however, makes good enough boxes, and leather a good enough cover. A recess should be cut around the boxes at the spindle, and tallow shaved fine and packed into it, before the leather is nailed down; and a further protection to it from grit and dust will be to slip a piece of woollen stocking leg on above the tallow; this, being elastic, hugs the spindle and absorbs the grit.

The curb inclosing the millstones is generally made to clear the stone about an inch and a half all around: thus its inside diameter is about three inches more than the diameter of the stone; the height being sufficient to clear the top of the stone two or three inches. There are many different ways of making the curb, among which the following is perhaps the best:—

Take good inch boards of sufficient width to make the top of four pieces, without other seams; these for an ordinary sized stone would require to be about twenty inches wide. To lay out the top, strike a circle on the floor the same diameter as the stone, then strike another outside of this, say an inch and a half, for clearance; then another three-fourths of an inch, or seven-eighths, outside of this again, for the thickness of the sides, and another an inch and a half outside of that, for the projection of the top over the sides. Next divide the last circle into four equal parts, and draw lines from one of these points across through the centre mark to the other mark opposite; draw another line the same across this one, connecting the other two points; this will divide the circle into four equal portions. Now place a board on one of these divisions, the outer edge even with the last (largest) circle, and strike that same circle upon the board; cut the board off by the circle, and use it for a pattern by which to cut the three other pieces; joint the inner straight edges of these and lay them down upon these quarters, two upon the floor on opposite sides of the circle, the other on top and across the ends of these. They must be laid carefully to the outer circle, and divided exactly alike by the corner lines. Now mark by the straight edge of the upper two across the lower two, and also by the straight edge of the lower ones across the upper cnes, for they must be halved together to these marks. It may be advisable to mark the face sides now, as they have to be gauge-marked around the edges, and cut away, and then fitted together again in the same position; they may be glued together, but should be riveted at the inside; the outside, being nailed down upon the staves, cannot give way. This makes a light cover, and being double at the corners, and each piece bearing its whole length upon the sides, it is also strong; the rim put around inside also adds to the strength of the top. This is about five inches wide, and one and a half thick, square on the outer edge and bevelled off a little around the inner edge; it is generally made of six pieces, and halved and lapped at the ends for additional strength. To place this rim, lay the cover down with the inside up, and strike a circle upon it corresponding with the one which marks the inside of the staves in the draft upon the floor, and fasten it around to the inside of this mark.

The perpendicular sides of the curb are composed of staves, and to mark these out properly a pattern is made by laying a thin piece of board upon the draft on the floor, and drawing the scratches upon it, corresponding with the out and insides of the staves in this draft; now mark both ends by this pattern; hew off the surplus wood pretty close to these marks, with an axe, then dress the outside and edges by the plane, using another pattern with a projecting end to test the bevel of the edges and insure accuracy, then gauge the thickness all around and dress off the inside. These are sometimes dowelled together, but it is better to run a small groove in the edges and put in a tongue or feather, which makes a better joint; these are fastened at the top by nailing, first through into the inside rim, and then down through the cover. A thin moulding, about two inches wide, is afterwards bent around to cover the nails and make a finish. The lower end is secured and finished by shrinking on a round iron band of half an inch in diameter, about four inches from the end, and bending a piece of wood around to cover the space from the band to the end, similar to the upper one. The iron band in this case forms the "bead" around the edge of this moulding; the ends of these wooden bands should be spliced by a thin plate of iron set in and secured by screws; this makes a light strong curb, and if made of nice wood and tastefully stained and varnished, it looks well and is durable.

There are many fancy curbs made generally of circles and squares alternately, and resembling panel work; these are more or less double, with many seams, and empty spaces in which the dust lodges and accumulates, and soon makes them heavy to handle, besides the steam generated by the heat and damp grain, in the process of grinding, condenses in these cavities and induces fermentation, and the consequence is a speedy decay of the woodwork of the curb. We have seen some of these fancy curbs composed of wood and zinc, covered on top with oil cloth, that were so rotten in a few years, that they would scarcely bear to be lifted, and soon went to pieces. The other being thin and single, is easily swept and kept clean, and will last like other inside fixtures, until it meets with an accident.

The opening in the centre of the cover is about thirty inches across; it may be marked and cut to the circle after it is fastened on, and a thin moulding nailed round inside and projecting with a round edge above the cover to protect the corners, and prevent anything being swept in upon the stone.

The hopper frame is about four feet long and two wide, with a bar across each end, and one in the middle. In the centre of the middle bar is the bearing of the top of the dansil, and between this and the hind bar the hopper is placed; between this and the front bar, a roller is placed across, upon which the string supporting the front of the shoe is wound; by turning this roller one way, the string is slackened and the feed increased, and by turning it in the opposite direction the string is tightened and the feed diminished. The frame stands fourteen or fifteen inches above the curb, and embraces the hopper near the middle. It is supported upon four legs, two near each end; these are generally turned, although when made to match the fancy curbs referred to, they are scolloped out into a kind of scroll shape.

To locate the hopper frame in its place, set the curb in its proper position with the outlet on the meal spout, then set on the frame in the position it is to stand, and hang a plumb down through the centre of the dansil bearing in the centre-bar, and shift the frame until the point of the plumb-bob is in line with the centre of the cock head of the spindle; mark around the legs on the top of the curb and cut out the holes for them. The legs should have a shoulder to rest upon the curb, and a round tenon to slip into these holes.

The hopper is a troublesome article for an inexperienced hand to make, although it looks quite simple. The first one that a man might make without any directions from a practised hand would show him some *wrinkles* in angles and their combinations, that he probably never thought of. It may trouble us not a little to give any directions that will be sufficiently intelligible to be of much use. We tried years ago to find some *rule* of easy application, which could be remembered, to get the proper bevel to cut the boards by, and the proper one to dress the corner cleats to, instead of the intricate process in trigonometry by which we now obtain them, but failed to hit upon any that was perfectly accurate. The following rules are applicable, although they both require a little dressing off to complete the fit:—

To get the bevel to cut the side boards, strike a circle, and step the circumference at five times; draw lines from these five points to the centre, and any one of these divisions is the angle to cut the boards.

To get the bevel for the corner cleat, strike a small circle, and divide it in three; one of these is the bevel for the corner posts, nearly. We have sometimes made these by turning a suitable sized stick in the lathe and marking it out in three equal divisions, and slitting to the centre at these three marks. These have to be dressed and fitted to the corner.

The following directions are reliable, and will make an ordinary square hopper to hold about six bushels, which may be made larger or smaller by preserving the same proportions: To cut the side boards, measure off along the edge of the board (or a line representing it) three feet and six inches, and at the centre of this distance draw a perpendicular. Measure up this perpendicular two feet and four inches, and from that point draw a scratch to each end of the base line. These will mark one side of the hopper, the other three being the same. Three inches in width, on the perpendicular, may be left off from the point of each piece to make the hole in the bottom; and if a single board can be gotten wide enough to make each side of one piece, it is preferable; if they must be spliced, do it at the point. Mitre these together equally at the corners, and fit in the corner posts, which should be of good wood and no larger than actually necessary to hold the screws.

We would advise the setting of the bevel by the corner itself, as the best rule, then work the corner pieces by this bevel inside as well as out. It is easier than to round them by hand, and they are better. It will improve the appearance of the hopper, and diminish the liability of grain to be spilt over, to put a large rounded moulding around the upper edge.

Some millers prefer a round hopper. These resemble a reversed cone instead of a pyramid, and are made of tapering staves with the wide ends up, and the narrow ends meeting in a cast-iron ring to which the points are fastened; this ring forms the hole in the bottom. A round rim of felloes is put around the top like a moulding to which the staves are secured, and a similar one is fastened round the middle. A board cover is laid on the hopper stand with a circular hole in its centre, into which the hopper is dropped down to the centre rim upon which it rests. This allows the hopper, board and all, to be slid forward to increase the feed, or back to stop the feed when the stones are stopped, just as the square hopper can be slid upon the naked frame.

The ordinary shoe that is hung under the hopper to control the feed may be made as follows: Take a piece of hard wood about twenty-two inches long, five inches wide, and one and a half thick. Round off the front end tastefully and work a mortise through it five or six inches long, leaving two inches of whole wood at the round front end and an inch and a half at each side; this is for the dansil and grain to pass through. Now bevel both sides to the distance of fifteen inches from the back end, and also the back end to the same angle as the hopper, and nail sides, six or seven inches deep upon these bevel edges, and an end behind to correspond with these sides, the end being rounded up in the middle, and the sides rounded down in front. Slant off the bottom down to the lower edge at the hole, to encourage the flow of the grain and the shoe will be complete. The back end should be nicely poised upon a pivot, fastened to the top of the curb, or hung upon a strap of leather or iron attached to the hopper stand.

The last is the most convenient arrangement, as the shoe in this case is lifted off with the stand. In the other case, the person lifting the stand must either lift the shoe with one hand, or unfasten the string in front, and lift the shoe by itself. The front end of the shoe is hung, as already indicated, by a string wound around a roller placed across the hopper frame. This string is attached to the roller near one end, and by drawing obliquely, keeps the side of the shoe in contact with the revolving dansil, and produces the necessary shake to cause the grain to feed.

The shoe is sometimes made by nailing the two side boards together at the bottom at a right angle, and cutting one off before it reaches the dansil, to discharge the grain. A point of the other board extends past the dansil, and to this the feed cord is attached; the hind end of this is boarded up like the other. This is a simple way of making the shoe, and if a small piece is fitted in to fill the sharp corner at the bottom, it answers a good purpose. We have seen cast-iron shoes used, but would not recommend them. They are heavy, and produce a lumbering unpleasant noise when the mill is working.

We will next make a few observations on the mill-stones, as these are the most important parts of the mill. The quantity and quality of the work done depend very much upon the kind of stones selected, and the condition in which they are kept.

Formerly mill stones were made of granite, or some other flinty conglomerate rock. Not unfrequently from boulders found in the vicinity where the mill was built, and in those days the millwright and miller's trade in cluded the making of these. These stones are scarcely ever used at the present day, except in very remote new settlements, being almost entirely superseded by the French Behr, or "burr stones." This displacement, however, has been gradual, and one kind of these rock stones, known as Esopus or "Soper" (Yankee) stone, maintained its ground against the burr stone innovation until very recently, and a few of these still linger in good mills, in various parts of the country, to the present time. These stones resemble in texture a bed of small white pebbles congealed together in a darker matrix, which completely fills the interstices, and leaves no empty cells like those in the burr, the sharp edges of which perform so important a part in the process of grinding. The Esopus stone has therefore no cutting edge except the dress made by the mill pick, and when this is worn out, it takes a polish as smooth as glass, and to grind with it in this condition would be like trying to saw with the back of a saw. It is also much softer than the burr stone, and for this reason the dress is much sooner worn out; but this softer nature tells in its favor in one respect, that it is far easier to dress than the burr stone, and this fact helps to reconcile millers to its use.

The last Esopus stones that we used were placed for bed-stones under very open burr runners. We expected that the harder burr stone would switch the dress off from these bed-stones in a short time, but such was not the case. The close soper stone had so much more wearing face than the open burr, that they wore nearly equally.

The following method of using rock and burr stone together we have heard highly recommended: Make the eye and a portion surrounding it of good quality of rock stone, and of such diameter that one tier of burr blocks put around the outside will make the stone of the intended size, hoop and finish up like ordinary burr stones. It is claimed in recommendation of this plan, that the flouring is all done near the outer skirt of the stone, and the portion around the eye, having only to crack the kernels, without coming in contact with the opposite stone, requires frequent dressing down with the pick to maintain the necessary clearance or "bosom" as it is called. Also, that this rock stone around the eye is so much softer than the burr stone at the skirt, that they will wear down nearly equally, and little more is required to keep the stones in order, than to keep the furrows in shape, and the dress sharp. But the making of mill-stones is now a separate branch of business, and as such it is brought to greater perfection than could have been expected had it continued to be one of the millwright's many responsibilities.

This consideration will render a chapter on making

mill-stones unnecessary, which in our younger days would have been as interesting to the millwright as any other in the book.

The millwright, until recently, had often to do all the turning, boring, and fitting for the whole mill, which is now done in the finishing shops, and is another important relief.

In selecting new stones, particular attention must be paid to the kind of grain intended to be ground, as the texture of the burr most suitable for one particular kind of grain, would be found totally unfit for other varieties. For some hard, dry, flinty varieties of wheat, such as the "Black Sea," a very close compact stone is required to reduce the grain sufficiently soft and fine without cutting the bran up too much. Such stones, unless very carefully kept, would be liable to heat and kill other soft varieties, particularly of winter wheat, and would neither grind such wheat so fast nor so well as a more open and a sharper variety of burr. Again, for corn and other coarse grains, a moderately open stone will grind faster and easier than such a close one, besides being much less difficult to keep sharp. On the other hand, we have often seen stones so open in texture, and with so little good face that it was impossible to grind these coarse grains sufficiently equal. And at the same time these very open stones, almost resembling a honey comb, would grind oats, and barley, and mixed grain, sufficiently fine for feed, as fast as it could be run into the stones and got away again, and rasping up the hulls among the meal, in a manner quite impossible with close-grained stones.

New stones should therefore be chosen with a due regard to the kind of grain they are intended to grind, bearing in mind that it is desirable to have the whole face of the stone as nearly alike as possible, as every large hole is so much lost to the face, and every perfectly close spot is also a loss, because there are no cutting edges in it except such as are made by the mill-pick, which is a very tedious and expensive process. If all the time that has been spent in work with the mill-pick, upon an old run of close burr stones, were counted and figured up at a moderate price, the sum would buy new stones a great many times over. On the other hand, we know of several runs of the honey-comb kind that are worn out, or nearly so, that never cost an hour's work in sharpening or dressing. All that they ever require being an overhauling once in several months, to furrow them out and true the face a little.

The only rule that we could give for selecting a run of burr stones, aside from their adaptation to a particular kind of grain, would be to get them with as many cells, and these as small as possible, and nearly equal throughout, any excess of cells being more admissible, or less damage in the runner, than in the bed-stone.

With regard to facing and dressing the new stones, we would observe that all new burr stones require a considerable depth taken off the face after they are put together, to fit them for taking the proper dress. One reason for this is, that the workman in facing and fitting the blocks must strike pretty heavy blows in order to make reasonable progress with his work. These heavy strokes fracture and splinter the surface considerably, below the portion actually taken off. Besides, in dressing off the edge to fit the blocks together, this hard splintery stone cannot be worked up to a true straight corner next the face, like a piece of wood or metal. But after the blocks are together, and the cement between, the corners are supported, and will bear to be worked down true like the rest of the stone, by a sharp light mill-pick. For these reasons the whole face of the stone must be picked off to a considerable depth with the pick; this is generally done, in part at least, at the factory where the stones are made, and to complete the face, and take it out of wind, the following is the common practice:—

Procure two straight edges parallel and straight on *both* edges and of exactly the same width. Place these, one on each side of the stone parallel with each other, and half way from the eye to the outside, mark along the sides of both of these, with a pencil or red chalk, sight across the top of these to see which end requires to be most settled into the stone to bring them true on top, then lift them, and go to work with the pick between the pencil marks, until you settle these marked portions to a good face and true with each other. After picking down roughly to near the depth, paint the straight edge and rub it lengthwise in the track upon the stone, to show the high spots and pick these down, until the face in these two channels is completed.

Now place the two paint staffs across the stone in the opposite direction, and in the same position, and mark out two other channels across the first, and settle these down with the pick, the same way until the paint shows on the crossing. Next place these across the corners and mark and pick down as before, until these channels meet the others, and so continue to mark and pick off, until the superfluous stone is removed to the proper depth, using the first four channels for a guide; and then rub the paint staff lightly over the whole surface and pick off the marked spots until the face is completed.

Some use three parallel sticks and place them in a triangle to sight across, and take the face out of wind, but two are preferable. We have sometimes faced new stones by a machine called a tram. This is a heavy paint staff of hard wood, hung at the middle upon a true turned standard fastened firmly at the centre of the eye, and raised or lowered by a thumb screw at the top. This was painted on the lower side, and slipped on to the standard, and the height gauged by the thumb-screw, so that the paint would just touch the highest points on the face of the stone. These marked points were cut down by the pick, and the tram lowered by the thumb-screw until it marked again; this process was continued until the tram swept the face of the stone all around.

This plan answers very well when the stone is nearly true, and requires but little taken off, but when there is considerable to be taken off, the process is too slow, as the whole surplus stone must be picked into dust, by going over and over the same ground, perhaps a hundred times; while by the other process the depth to be cut away is shown at first, and can be mostly knocked off in larger chips and splinters almost to the full depth at once.

It will facilitate the reduction of the face to a proper finish to rub the surface over with a burr block, or other hard stone with a good face, to smooth down the sharp points, which will otherwise keep up the paint stick and tend to scratch and spoil its surface.

Some put the stones down in this condition and grind them together to finish and fit the faces, and then take them up and furrow and dress them for grinding. Others furrow them out first roughly, and then grind and fit the faces, and take them up and finish the furrows, and crack and dress for grinding. By the last plan, less grinding will suffice, because the furrows being cut out it leaves only the lands between to be ground down, instead of the whole surface of the stone; but, on the other hand, we have frequently seen the furrows carefully marked and dug out before the stones were ground together, which were either almost or altogether ground out and obliterated, and had to be marked and made anew. We are not disposed to argue this point, and will merely remark that the process of grinding down the face is slow and tedious, if the naked faces of the stones are ground in contact, but by interposing a small stream of *dry* sand, or *clean* water it is done more quickly and better. Never try *wet* sand or sand and water. We were once persuaded to use this, but the friction and heat reduced the wet sand into clay or mud, and packed every crevice in the stones full, and everything around was daubed full and plastered with it.

When dry sand, or clean water is used it is easily controlled. You may place a belt on edge around and near the stone, and, if sand is used, collect it inside of this, and keep pouring it into the eye again. If water, fill the eye around the spindle with mill dust to keep the bush dry, and lay a ridge of similar dust around inside of the belt, and make one or more little spouts from this to discharge the water, which may either be caught under these in a vessel or run away. The water may be poured into the same spot in the eye from a tea-kettle, a piece of rag being laid under the small stream to save the dust from washing. If no belt or other convenience can be had to place around the stone, then the curb must be put on, for without some kind of fence the sand or water would be scattered all over the mill.

The stones having been brought to a proper face, the furrows are next to be laid out. To do this, fit a piece of board into the eye level with the face of the stone, and find the centre by moving a sweep around the skirt, and from this centre draw a circle upon the board with a radius equal to the intended draught of the furrows. Step the circumference of the stone into as many divisions as there are to be quarters in the dress, and mark each of these divisions on the hoop of the stone with a cold chisel. Now lay a furrow pattern from each of these marks on the hoop to the *outside* of the draught circle, on the board, and mark the furrows on the face of the stone by this pattern.

These will indicate the leading furrows, and to mark the short furrows place another pattern representing the land, along side of the leading furrows, and the pattern for the furrow beyond that again, and so on alternately, marking the furrow and the land each time until the whole quarter is laid out.

We give an engraving of a stone with a dress of eighteen quarters or sections, and also one of a stone with a dress of twenty-one quarters or sections, which we think will be sufficient to illustrate, and assist in explaining the remarks we have already made and those which are to follow on this subject.

The rule for laying out the dress of mill-stones, as above stated, will be understood at once, by every miller, but to the uninitiated it will be incomprehensible without the following explanation :—

When the grain falls into the eye of the stone it is swept around by the revolving irons until caught between the stones, and here the process of grinding commences. At first the grain is only cracked, and a motion given to it, which increases with every revolution as it is driven around between the stones, and reduced finer and finer as it approaches the circumference where it is discharged as flour or meal. This velocity, and the ratio of its increase, can neither be measured nor computed by any

Fig. 38.

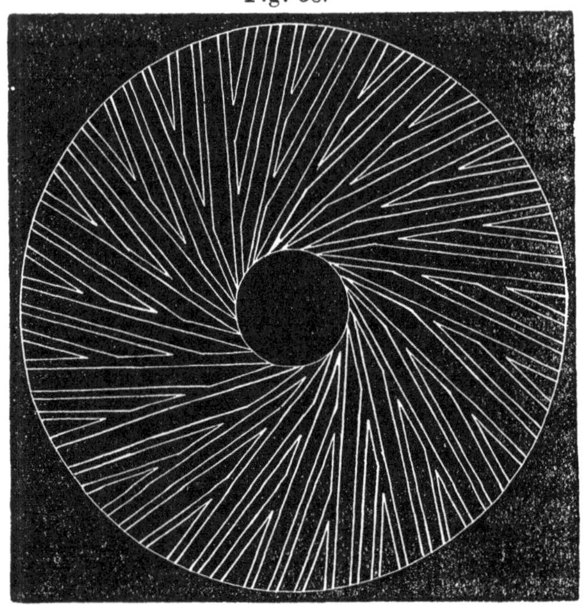

Fig. 39.

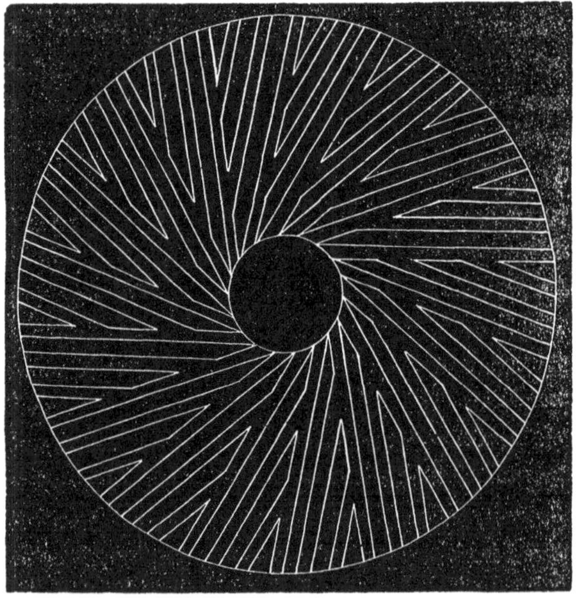

known process, but as it is urged onward by the motion of the runner, and retarded by the quiescent bed-stone, its revolving velocity at every point is somewhere between these two extremes. Its progress towards the circumference, by the centrifugal force, is also modified by these two contingencies, and the problem to be solved is, to proportion the draught of the furrows, and the bosom and dress of the stones, so that the grain in travelling from the eye to the circumference, in this complicated and undefinable course, will be thoroughly ground, and no more, when it comes through.

It has been already mentioned, that if too long confined, the flour will be heated and killed, so that it will not raise and make bread; on the other hand, if it pass through too rapidly it will not be sufficiently ground, and part will pass over with the bran, and cause a waste.

We have also said that this subject of furrows and their draft was understood by every miller; this will seem incompatible with the vague uncertainty expressed in these last paragraphs, and be received with some modification, as it is the *result* and not the *principle* that is understood. And even *results* are very vaguely and differently understood by different millers.

Here is an example:—We once heard two first-class millers discussing the propriety of using an Esopus bed-stone under a burr runner. The advocate for this arrangement stated with perfect confidence that the running stone performed two-thirds of the work in grinding. The opponent disputed the allegation, and challenged its proof, at the same time remarking that it was impossible to get between the stones to see which did the most work, but this much he knew, that when a run of stones, both alike in quality, was dressed alike, the dress in the bed-stone would be worn out first. This observation we have often seen verified since, but the theoretical demonstration as to its cause we leave to philosophers.

Again, we employed a stone-dresser, reputed to be a first-rate workman, and he deserved the reputation so far as the mechanical part was concerned, who took great pains to impress the younger workmen with the importance of cutting the *deep* edge of the furrows steep and sharp, because, he said, it was these sharp edges that broke and cut the grain. How any man, of good judgment otherwise, could nourish such a fallacy through so many years of experience, is a mystery not easily explained, especially as he must have seen these steep corners which he advocated, invariably filled up with dust, to an angle less than half as steep as that which he insisted upon.

We mention these differences and difficulties to show the uncertain and variable data upon which this problem depends, and instead of offering any particular rule as applicable under all circumstances, we will only give some general observations on the most popular dresses and draught in use, and advise any miller to depend rather upon his own judgment and experience than anything we can suggest, to determine the dress most suitable for the particular case in hand.

The dress most popular in our younger days was a quarter dress, consisting of either seven or nine quarters; the rule being to give the leading furrows an inch of draught for every foot in diameter of the stone, but this reminds us that we undertook to explain the meaning of this *draught*, and have failed to fulfil the promise.

The furrows do not run from the skirt of the stone direct to the centre, but obliquely to one side, and the distance of the inner end from the centre is the measure of the draught. These furrows point to the right or left of the centre, according to the direction in which the stone is to revolve, whether with, or against the sun, the object being to sweep the grain outward from the eye, and assist its progress through the stone. The dress is laid on both stones alike, with the face of both up, and when the runner is *reversed* and laid upon the bed-stone, the furrows cross each other at an angle, like the blades of a pair of shears while closing. But although the closing or crossing of the furrows, like that of the shears, commences at the centre and progresses outward, they never come together, like the shears, but every furrow in the runner is constantly sweeping across the corresponding furrows of the bed-stone, with this closing progressive motion, shearing the grain finer between the inclined planes of the furrows, and carrying it over the intervening lands, the sharp and even faces of which complete the reduction of the particles to the proper grade, and finish the process of grinding.

We knew a man, the owner of a mill (his own mill-wright and miller, said to be very ingenious, and a real "jack of all trades"), who changed the breast-wheel which drove his mill into an overshot; this reversed the motion of the machinery, and turned the stones in the opposite direction. To adapt the stones to this revolution he altered the *shape* of the furrows, by deepening the inner edge, and turning the feather edge of the inclined plane to the opposite side, but left the *draught* and position of the furrows the same as before. When the stones were started the furrows gathered the grain toward the centre, instead of sweeping it outward, and thus counteracted the centrifugal force, instead of helping it; and the flour, instead of discharging as it was ground, was retained between the stones until it heated and was baked on to the surface, clogging and stopping the motion. The stones were raised to free them, but the flour was baked on more and more, until they were taken up and the paste picked off with the mill-picks. He tried them several times, but always with the same result, the trials occupying several days, for it was a tedious job to pick the paste out of the furrows. The man was very superstitious, and as there were several reputed witches in the neighborhood, he charged them with trying to annoy him.

We were engaged with a gang of hands on a mill eight miles distant, and he came over to see if we or any of our men could help him out of his dilemma. We inquired if he had changed the furrows in the stones, he said that he had, and we failed to make out by his account what the trouble was, and sent a miller over with him to reconnoitre, and he found the stones as we have described, and pointed out the remedy.

We mention this circumstance for the double purpose of illustrating the importance of this subject of draught, and also to verify a remark that we have often had occasion to make, which is, that a man to make a good and successful millwright, or even miller, must have more than an ordinary share of judgment and common sense; and many that would make first-class mechanics in almost any ordinary trade, would, if they choose this occupation, often meet with circumstances and unforeseen occurrences that would tax their patience and ingenuity to the very utmost, and be a frequent source of perplexity and annoyance.

The dress we began to describe as consisting of only seven or nine quarters, had of course only that number of leading furrows meeting in the eye, to admit the grain between the stones, but this deficiency was balanced by the *bosom*, which at that time was always given to the running stone. The bed-stone was faced true and level throughout, as at present, but the runner, after being faced true by a *real straight edge*, was then gone over with another that had a swell at the eye of from an eighth to a quarter of an inch, and ran out to nothing at a certain distance from the skirt; this had a hole through the middle, and was slipped on to a pin in the centre of the eye to keep it true, and being painted below and turned round upon the centre-pin, the paint marked the stone around the eye, which was picked off, and the process repeated until the bosom was excavated to the shape of the swelled straight edge or mould.

This open bosom admitted the grain between the stones as freely as the increased number of furrows now generally used, with little or no bosom, the grain at its entrance being only cracked, and then reduced gradually as the bosom contracted to its termination. From this point to the skirt, both stones were perfectly true, and this was called the flour or face of the stones, it being the part where the actual *flouring* was performed, and the only portion of the faces that could come in contact. This flouring face was left of various widths, according to the fancy or experience of different millers, and modified to suit the kind of grain and the quality of the stones. Sometimes it extended only five inches from the skirt, while under other circumstances it continued half way to the eye.

Since that time the bosom has been more or less discarded, the number of quarters and leading furrows being proportionately increased. Another modification has also taken place, which is, that the short furrows are now generally cut through into the leaders, making an open communication with these, which greatly increases the draught. Formerly a whole *land* intervened between, against which the short furrows butted, and over which the grain had to pass to get into them. This left a whole unbroken land on both sides of the leading furrows, and made some kind of bosom indispensable. In order to understand this subject perfectly, the reader must study the drawings, for without having recourse to these we cannot convey our ideas intelligibly.

Some enterprising miller ventured to increase the number of quarters from nine to eleven; another with still more self-reliance raised them to thirteen; then another bolder than either got them up to fifteen; the next one to seventeen, and so on, still preferring the odd numbers, until twenty-one was the favorite number. During the progress of this improvement, the short furrows were gradually cut through into the leaders, until that practice is almost universal; and the bosom has been reduced, generally to the thickness of a sheet of paper, and by some good millers it is dispensed with altogether. The most popular rule, and probably the best one, is, to place a sheet of writing paper under the straight paint-staff, and have it draw freely out all around the eye.

While these transitions were progressing, the ratio of the draught to the diameter of the stone was frequently modified; when the quarters were about fifteen, and the short furrows cut through into the leaders, an inch to the foot was found to be too much, but when the whole land was intervened, this draught was not sufficient. Up to about this number, and above this, the furrows are generally cut through. The necessity for these variations will be understood by consulting the cuts; the fewer leaders, the more short furrows must intervene, to fill the quarter, and the draught of every one of these increases as they increase in number. The actual obstruction interposed by the whole land between will also be seen. The average draught now used with the increased number of leading furrows, and the short ones cut through, may be said to be the old rule of an inch to the foot.

One cause for these alterations in the dress of millstones may be found in the increased power now applied, and the greater quantity now ground in a given time by each run. That this was the principal cause will be still more apparent, when we remember that the *size* of the stones now used to grind this greater quantity is much smaller than those formerly used to grind the less quantity. The stones formerly used varied in size from four and a half up to six feet; those now in use vary from four and a half *down* to sizes ridiculously small for the quantity of work expected of them. A five foot stone could afford to have a part of its face wasted in bosom, and a little more wasted in extra wide lands, and yet grind five or six bushels in an hour; but when a four foot stone has to grind ten or twenty bushels in the same time, there is little margin for waste, and both the miller and millwright are put upon their mettle to meet the altered contingency.

This subject reminds us of an idea prevalent among the uninitiated with which we have often been "bored," viz., that where the power is limited, small stones should be used to economize that power. This is a fallacy, a certain number of superficial feet of face must pass each other in order to grate and grind a given quantity, whether the stone be large or small. If the stones be large, fewer revolutions will pass this number of feet, and the driving pulley or pinion on the spindle will be large and have a strong purchase to turn it; but if the stones are small, it will take more revolutions to do it, and the pinion or pulley must be proportionately smaller, and have less purchase. If the small stones be half the superficial face of the large ones, then they must make twice as many revolutions, and with only half the purchase of the large ones; this would make them equal, but the small ones have a decided disadvantage in having to perform the same amount of work upon one-half of the surface. All the economy then that we can see in using small stones is, that they cost less, occupy less room, and are more easily moved and handled.

There is another kind of dress used to a considerable extent in some localities, called the "sickle" dress, or circle dress. This, as the name would indicate, is composed of a series of circles, resembling the reaping sickle, instead of a combination of straight lines like the other. These circles commence in the eye and extend to the circumference in a peculiar curve, corresponding with the draught of the other leading furrows, and balancing the angle of the short ones in the quarter dress, a single short furrow, agreeing with these, is generally intervened to divide the space between at the outer ends. This dress, when carefully laid and worked out, gives a regular and equal draught throughout its whole length, and considered theoretically, it is the most perfect dress.

But there are certain practical objections to it, which have hitherto prevented it from coming into general use, and tend to give the quarter dress the preference. One is, that having so many furrows crowded into the eye, these inner ends must be the smallest, and being only half the number, and the whole grain having to pass through them in this limited space, they ought, like the leaders in the quarter dress, to be largest next the inner end. Another is, that the varying curve of this dress prevents the use of the painted furrow patterns to true and finish the furrows and keep them of the proper size and shape.

This is a serious objection, although those millers who only dress out their furrows once or twice a year may not consider it so. This practice is far too general, and is like that of a sawyer who would keep his saws filed and sharp, but would only set them once or twice a year, and at these setting times give them an extra allowance to help them over the long interval; this policy of the sawyer would be absurd, but no more so than that of the miller in the parallel case. The true policy for the sawyer is, when he files the saw, to examine the set, and spread a point here, and bend a tooth there, and thus keep the set true and sufficient; the true policy for the miller is, when he picks the stones to rub the painted pattern in the furrows, and pick off the marks left, deepening a spot here and widening a spot there, and thus *keeping* the furrows true and sufficient.

This is even more imperative with the stone than with the saw, because it is the feather edges of the furrows that break and reduce the particles of grain, therefore these edges soon wear smooth, while it is important that they should be kept sharp. The system which we advocate insures this, as these feather edges are more or less sharpened every time the stones are picked, while by the other system they soon become smooth as glass and are allowed to continue so until the great furrowing out time (like Aunt Dinah's great "claring up time") comes round.

It has been already intimated that the sickle dress does not admit of this plan of keeping the furrows, and this has contributed more than any other cause to inaugurate the system which we deprecate, of leaving the furrows untouched until worn shallow and smooth, and then dig them out again.

Some make another objection to the sickle dress, of the apparent difficulty of tracing the proper curve. This alone should not deter any one from using it. The rule in trigonometry by which it is arrived at is rather intricate, and would hardly be understood by one miller in a hundred, if we should give it here; but any one may get it from another miller, or from the patterns used in some other mill, or failing these sources, let him lay out two leading furrows of the quarter dress upon the new stone, or on the floor, and fill up the intermediate space with lands and short furrows to match. Now take a suitable piece of thin board and lay upon this quarter, and trace the intended curve upon it, by balancing the draught of the leader, and the angle of the different short furrows, and splitting the difference as nearly as possible by the eye. Cut off the outer edge of the board to this curve, and then test and correct it by the following process: Lay the curve thus made upon a clean part of the stone (or floor), the inner end with the intended draught, and mark it plainly; now reverse the curve the other side up, and with the draught right to represent a *furrow of the runner*, and move it slowly across the curve on the bed-stone, like an opposite furrow, and measure the angle at which they cross all the way from the eye to the circumference.

If the angle between the round sides as they close like a pair of shears all the way out be the same, the curve is right and the draught equal; if the angle varies, the curve must be altered until the above result be attained. This curve will now be the outside of the required pattern; to finish it, mark or gauge from this side the intended width of the furrow upon the board, and cut it away to the mark, and the pattern will be finished, except that a piece must either be left on or attached to the *inner* side at the *inner* end, wide enough to mark the draught distance upon, and a hole bored through it exactly at the centre of the stone to be slipped on to a pin at that spot to keep the pattern true. Every leading furrow pattern requires this pivot, and to lay out the furrows it is only necessary to move the outer end from one mark to the next one on the hoop to lay out the divisions, the pivot keeping the inner end always right.

There is another dress which is a sort of compromise between the quarter dress and the sickle. It proceeds in a straight course from the eye to within six inches of the skirt, and then turns a corner, resembling the short furrows of the quarter dress from this point. It was, no doubt, intended to combine all the good points of both the others, but fails to attain those of either. It has neither the even regular draught of the sickle dress, nor the straight freedom of furrows of the other, while it combines the disadvantages of both.

The actual width of the furrows and lands, and the relative proportion of these to each other, is another point equally vague. We have seen furrows of every intermediate width from three-quarters of an inch up to two inches, and lands varying from three-fourths of an inch to four inches, and these jumbled together in every conceivable proportion, without any resemblance to a rule to regulate the relative breadth of each to the other. The greatest curiosity in this direction is this, that each individual miller considers his own system right, and every other miller to be more or less in error in proportion as they differ from him, and if a man were to assail his system and try to persuade him to change it, it would be like trying to make a proselyte in religion. All the benefit likely to result in either case would be from a casual random remark that might set him to *thinking*.

These considerations, and the great length to which this article has already expanded, have tired us of the subject, and as the reader must be equally tired, and we have yet to consider the subject of balancing the running stone, we will appeal to Mr. Littlejohn to help us through this part of the subject. We saw an article from him published some years ago in the "Scientific American," which expressed our ideas and experience so much better than we could, that we will transcribe a portion of it for the benefit of our readers.

It may be remarked, however, that the breadth of the land is now generally made to approach much nearer to that of the furrow than formerly, some good millers making them equal, that is, half furrow and half land, while others, like Mr. Littlejohn, make the furrows the widest. But this reduction of the land must be modified by the quality of the stone, and the kind and condition of the grain expected to be ground. The "dress" cracked into the lands by the mill-pick to give the requisite sharp cutting surface, is also modified by these considerations, and still more by the fancies and capacities of the different millers. These combined give it a range from four cracks to the inch, up to about thirty-two; the best plan is to have them as close as will leave each one distinct from the next, with an unbroken line of face between.

The process of balancing the running-stone by the bale upon the pivot of the spindle is now, by the improvements introduced by the makers, easily accomplished. This pivot, or point of suspension, is near the *average* height of the burr blocks composing the face of the stone, but these blocks are of very different thicknesses and weights, and formerly no attention was paid to this difference; the result was that the burr portion of the stone below the point of suspension was frequently much heavier and thicker on one side than the other. Then in backing up the stone to the full thickness, large blocks of stone were frequently placed in, over the thin side; these were above the point of suspension, and being heavier than the plaster which filled the opposite side, the stone would be heavy at one side next the back, and at the other next the face, and thus it was easy to balance it when at rest; but when set in motion, the heavy side next the face would be drawn in that direction by its centrifugal force, while the opposite side, being heavy next the back, would be drawn in the other direction. These centrifugal forces, being equal and opposite, one above and the other below the point of suspension, or top of the spindle, the stone would be tipped over, and one side would drag upon the bed-stone whenever it was set in motion. This drag would increase in weight as the velocity increased, and diminish again with the velocity, until with it, it finally subsided.

This condition of the running-stone has caused much perplexity and loss, on account of the difficulty of determining the precise spot to place the lead to balance it. It would be easy to give it a running balance by placing the lead inside the hoop at the top, on the high side, but this would disturb the standing balance, and that side would drag every time both in stopping and starting. Unless a stone be balanced in both conditions, it is impossible to keep it in perfect order to do first-class work, although many good stones are used and worn out that never were perfectly balanced in both of these states.

The best means employed to attain both of these balances, before the introduction of Mr. Brown's system, was, to dig out either three or four large holes, at opposite sides, through the backing of the stone down to the burr.blocks, and close to the hoop. In each of these place a sheet iron canister the whole depth, and fitting to the hoop outside, the other sides of any shape; place an iron spindle, screwed the whole length, down the centre of each canister, and having free bearing in the bottom and top, with some provisions for turning it at top, and on this spindle place a heavy chunk of lead, fitting the canister around the sides, but free to screw up or down by turning the spindle. Particular care must be taken in building these in, to obtain the standing balance. This is attained by placing stone or iron around the boxes to fill vacancies on the light side of the stone, and pieces of wood, or, in extreme cases, vacancies covered by wood, in the heavy side, filling and covering the whole with plaster.

The stone may now be set in motion to ascertain the condition of the running balance. When the heavy and light sides are determined and marked, the stone may be stopped and the balance adjusted by screwing up the lead weights at the light side, and letting them down at the heavy side; these trials must be repeated, and the weights manipulated until the proper running balance is attained. This raising and lowering of the lead weights will not disturb the *standing* balance, but should it vary, as it is likely to do by the drying of the new plaster at first, or by wearing unequally afterwards, it can be readjusted by adding to or diminishing the weight of the lead at the high or low sides. When this occurs, of course the lead weights must be regulated anew to re-establish the running balance.

This mode of balancing a stone involves considerable expense of time and trouble. We have frequently had to dig out whole blocks of cull burr stone in making the holes for these canisters; at other times we have encountered large slabs of building stones, and in some cases it would be easier to remove the whole backing, hoops and all, and back the stone anew. But it is better policy for the miller to spend any amount of time and patience upon it and get it right, than to run it with one side constantly trailing upon the bed-stone.

The following method is that referred to as being introduced by Mr. Brown. We have used it for several years, and find it as reliable and much more expeditious, than the last described: Get two pieces of thin close-grained board about four feet long, and six inches wide; plane these and gauge and dress them down to about three-eighths of an inch throughout, raise the runner about half an inch clear of the bed-stone, and slip these pieces between, one at each side and about half way from the spindle to the outer edge; slip a piece under each projecting end of these to fill the space between them and the floor, and drive a nail *down* through to keep them all firm. Now set the stone in motion, and let it down until it scrubs upon the boards, and while it is fitting and polishing the surface of these, fit a plank across the top of the stone in a convenient position for a rest, and turn the whole back off perfectly true. This, if the face be kept down tight to the boards during the operation, will make the back agree exactly with the face, and this is the main point gained by Brown's system, as will be seen by the other part of the operation.

The stone may now be stopped and raised up and the boards removed, and when started again at the working velocity and clear of the bed-stone, it will assume its regular running position, and the light side may be marked by holding a lead pencil against the rest plank, and moving it carefully down until it touches the new turned back of the stone.

The stone being stopped, a portion of the plaster may be dug out of the back next the hoop, at the side marked by the pencil, and the space filled with lead. It may now be started again, and the result tested by the pencil, as before, and lead put in the same way next the new pencil mark. This must be continued until the pencil will mark equally all around the back, when the stone may be let down to its place, and as the face agrees exactly with the new turned back, it will run true with the bed-stone as long as the motion is kept up; but when the motion subsides, one side of the stone will drag upon the bed-stone, unless it be also at the true standing balance, which is very unlikely.

The problem now to be solved is, how to obtain the standing balance without disturbing the running one? We have already mentioned the pivot in the top of the spindle, as the point of suspension, and intimated that an extra weight in any part of the stone, above or below this point, would disturb the running balance by its centrifugal force; but we did not mention (unless by implication) that a weight added exactly on a level with that point, could not disturb the running balance in any way. This, however, is a fact, and this fact furnishes a solution of the problem.

The following is the mode of operation: Raise the stone clear of the bed-stone, and with a hand on each side move it each way until it is clear of the driver, and free to balance every way upon the pivot; now try it all round until you discover the light side, and place weights upon that side, until it balances alike. Mark the side across the hoop, and turn the stone on edge, with that side up, and measure the distance from the face of the stone to the pivot socket in the bale, then measure the same distance from the face on the edge of the stone, and mark that upon the hoop. The intersection of these two marks will designate the *centre* of the spot where the lead weight should be put in *theoretically*, but *practically* this is near the upper edge of the hole, instead of the centre, as the lead must be placed *two inches below* the point designated. The reason of this is, that the driving force is applied at an average of four inches below the point of suspension, and as the centrifugal force is affected by this, the neutral point is a medium between these. It may be difficult to explain this philosophically, but our own experience corroborates that of Mr. Brown, and establishes the fact beyond a question.

To complete the operation, determine the number of pounds in the weights required to balance, and take the same weight of lead and estimate the size of the hole required to hold it when melted and run in solid; lay out the hole at the part indicated, of an oblong form, if it must be large, and chisel it out the proper size; add more lead to balance the weight of hoop, stone, and plaster excavated, and melt the whole and pour it in. This will complete the standing balance. We have sometimes found it expedient to extend the lead outside of the hoop with rivets, but this is only necessary when the quantity required is very great.

It may be useful, in this connection, to mention some peculiarities with respect to the plaster used in these operations. In balancing an old stone by the last method, it is frequently necessary to cover the back anew, either in part or wholly with new plaster; this, as well as covering the lead or other holes in the backing, may be done by mixing the calcined plaster with clean water. Everything must be convenient, and much expedition should be used in applying it, or otherwise a considerable quantity will be wasted. The old plaster must be wet before the new is applied, and to make the adhesion certain, bit holes may be bored a short depth into the old surface, and where holes and angles occur nails should be driven in various directions, taking care to have the heads below the surface of the new plaster when it is applied.

To build in the canisters used in the other method of balancing, more care and time are required in order to arrive at the proper balance, and for this reason some kind of sizing must be used in the water, as this will prevent the plaster setting so fast. Common glue or isinglass, or failing these, milk to the water may be employed. Many never use sizing of any kind, but mix the plaster with urine instead of water, and this answers well. Alum mixed in the plaster makes it close and hard; some use this in the last coat of backing. A better way is to wash the newly-finished back with a strong solution of alum water several times in succession; cold water will not hold much alum in solution, but hot water will hold sufficient to make the plaster polish almost like marble.

It is sometimes difficult to obtain the ordinary calcined gypsum of commerce, and as only a small quantity is required for any of these operations, it may be prepared from the ordinary plaster of Paris used upon land. To do this, place a small quantity in a clean dry kettle upon the stove and boil it until it settles and lays still. The kettle must not be full, because it boils like soap, or sugar, and would be liable to run over; it should be stirred occasionally until sufficiently cooked, when it will settle down and remain at rest.

The stones referred to as so difficult to balance were built without any regard to balancing the different strata, or courses, of which the stone was composed. Sometimes they were built on a solid plank platform of the size and shape of the stone, but a more popular way was to build them on edge, setting the face by a straight-edge, and the skirt by a sweep from the centre. This mode was found convenient in fitting, as the seam could be seen on both sides, or through from one side to the other: it was also convenient in cementing, as the weight of the block was a convenient help to press and retain it in place while the cement was hardening, and for this reason the stone was rolled around as the blocks were fitted on, to keep the vacant side always up.

The process of making burr stones, as has been intimated, has undergone a decided improvement, and is conducted on a more enlightened principle than formerly. They are now made upon a true turned and adjusted cast-iron plate, this plate being itself balanced upon a pivot at the centre, like the running mill-stone. The blocks are first fitted upon this plate, and equipoised every way, and then each one of the strata of which the filling and back are composed, balanced as they are put on; thus every different course is balanced by itself, and the whole stone fully adjusted before it leaves the stocks, and all the further balancing it may require when hung upon the bale will only be so much as the socket in the bale might vary from that of the cast plate upon which it was built, or so much as the miller might take off one side more than another in dressing. And for the adjustment of any little variation from these causes, three or four small cast boxes are placed inside the hoop at equal distances, and having openings through the hoop, which are opened and closed by a screw. A little shot (lead) is introduced into these boxes to perfect the balance, which can be altered or regulated at any time by removing a little of this shot from one of these boxes, to another with a teaspoon.

The process of *bolting* may now be considered, it being the last operation in manufacturing the grain into flour, and that by which the different qualities of flour are separated from the bran and shorts. Within the past twenty-five years this has undergone greater improvement than even the process of grinding.

The bolt consists of a long reel covered with some kind of sieve, or bolting cloth, the texture of which is fine next the head where the flour enters, but coarser towards the tail end, where the meshes are wide enough to pass the shorts through, leaving the bran to pass through the whole length and be discharged at the lower end. The reel thus constructed forms a kind of cylinder, which is hung and made to revolve horizontally upon a bearing at each end; that supporting the head where the flour enters being about six inches higher than the one at the tail end where the bran escapes. It is this inclination of the reel which causes the contents to progress towards the tail end as it revolves, and by the sliding and tumbling it undergoes during this progression, the flour is sifted through. The finer portions being the heaviest and consequently next to the cloth find their way through first, and fall near the head, while the coarser fall next, and as the particles remaining are coarser and lighter, they pass through further and further down until near the tail end they approach the grade and gravity of bran, and find their way through the coarse cloth as "canaille," or shorts.

The oldest kind of bolt reel that we ever fitted up was covered with brass wire cloth; it was made to revolve pretty rapidly, and had another reel revolving slowly inside, the ribs of which were composed of brushes the whole length; these brushes swept the flour through the wire cloth and kept the meshes open. The reel was not six square like the modern one, but circular, and composed of two corresponding halves, which could be put together over the brushes, or taken apart conveniently. It was about six feet long and about two in diameter, and was composed of rims or semicircles fastened about six inches apart around the inside of longitudinal ribs which held them in place, and left the inside free and clear upon which to nail the wire cloth, and also allowed the hair brushes to revolve inside. One of these wire cloth and brush bolts (we have forgotten the technical name) would bolt for two or three run of stones.

Another, called the "English" or "Bag Bolt," was more extensively used than the wire cloth. The reel for this kind was about the same size as the other, but was made and operated on a very different principle. It was composed of six large round ribs placed lengthwise upon arms mortised through the shaft, like the modern reel, but the cloth was different, being both longer and wider than the reel, and woven throughout without a seam, of some kind of thread approaching silk in strength, but much coarser and firmer in texture; it was said at that time to be composed of the inner bark of some kind of nettle, but we have met with nothing since either to confirm or confute that opinion.

The coarse tail end of this bag without a bottom was

trimmed or bound with chamois leather, and had six loops which were slipped on to the ends of the ribs when it was drawn over the reel; the head end was bound with this leather much wider, and a noose made around the edge for a strong cord to pass through, by which to draw it together, leaving only a hole large enough to feed the grist. Being larger than the reel, and the whole driven at a great velocity, the cloth bulged out considerably beyond the diameter of the reel, and a round smooth bar was placed along each side against which the projecting cloth rubbed; this answered the same purpose as the brushes.

The bolt chest in which both of these kinds were inclosed were nearly alike, and made high enough to admit of collecting and conducting the two kinds of flour, and the shorts and bran by funnel-shaped spouts into separate bags. The grist did not generally pass directly from the stones to these bolts, but was elevated to an upper floor where it was spread and cooled, and afterwards shoved into another large hopper, from which it was fed into the bolt; this feeding was always attended with more or less trouble, from the tendency of the grist to pack and remain stationary. A peculiar kind of shoe was used for this purpose, more like a bellows than an ordinary grain shoe, being mostly composed of leather, and resembling an ordinary shoe in shape. (This similarity in shape and material was no doubt the origin of the name now given to this article.) It hung loosely behind by the leather of which it was composed, and the point which conducted the grist inside of the bolt cloth rested upon the shaft of the reel, which had iron strikers attached to shake the shoe and encourage the discharge. The shoe was large, and the velocity of the reel so great that the strikers rattled and banged at such a rate against the shoe that little else could be heard about the premises while the bolt was going.

Another machine was made to revolve over the cooling floor with a draft towards the centre, that collected the grist into this hopper, and stirred the contents to help it to feed, thus dispensing with the services of the boy whose business it was to attend to that part of the operation. This circumstance gave the name of "Hopper Boy" to a similar machine since used for cooling and collecting the bolted flour, in the modern packing machine.

The English bolt, like the Brush bolt, was capable of bolting fast enough for several run of stones, and even then it did not require to run all the time that the stones were grinding. They are scarcely ever used at the present day in the United States, although a few of them may still be found in some of the British provinces.

A first step toward the use of the present bolt and Dutch cloth was made by stretching the English bolting cloth upon the long modern reel, and tacking it to the reel as at present. Instead of the English bolt cloth, a kind of smooth hard book muslin was frequently used; to shake the flour through, and prevent it sticking to the cloth, a knocker was fixed upon the tail end journal, which raised and dropped that end of the reel, about an inch four times during each revolution; in addition to this, another knocker, and sometimes two were placed on top of the reel outside. This was a light piece of wood placed across the reel, the back end hinged upon a pin, leaving the front end free to move up and down; a piece of wood was fastened upon each of the six ribs of the reel under this striker to protect the bolt cloth, and as the reel revolved each of these pieces raised and dropped the striker as they passed under it; this kept up a continual pounding upon the ribs of the reel, at the same time that the dumper on the lower end kept raising and dropping that whole end, and between the two they managed to shake through the greater part of the flour. When several of these were in operation at the same time, they made considerable racket and noise. These precautions, moreover, did not prevent the cloth from being gradually filled and clogged up with dust and beards, and other accumulations which prevented the flour from getting freely through, and had occasionally to be removed. This cleansing of the bolt was attended with a good deal of trouble, especially in the muslin cloths, the texture of which was tender and the meshes easily displaced. The first resource was generally to brush it carefully with a stiff hair brush; this removed anything loose, with the dry dust, but the beards stuck through the cloth, and the dust more or less pasted into it by sweating or moisture and could not be removed by the brush, and it required considerable ingenuity to get rid of these.

One effectual, but tedious method of removing the beards was to shave the cloth all over with a razor; this cut the beards close to the outside of the cloth, and by rubbing it over with the hand or a brush, the inner ends dropped through to the inside. When this was carefully done by an expert hand, the cloth kept cleaner afterwards, as it shaved off all projecting hairs or nap from the outside, and there was always more or less of this upon these cloths. Another expedient was to fix a soft piece of chamois leather or India rubber in a suitable cushion or the palm of the hand, and sweep this dexterously over the cloth, bringing it across a piece ot sheepskin at each return stroke to remove the beards adhering to it. These beards are all toothed along the edges, the teeth, or barbs rather, all lying in one direction, like the barb of a fish-hook or spear. This arrangement of the barbs allows the beard to pass through anything freely with one particular end foremost, but never allows it to back up. We have seen many fowls killed by swallowing the beards of a new kind of wheat along with the grain; these worked their way through the crop in every direction, many of them coming out through the skin and feathers at the breast; it is this peculiarity that enables the chamois or rubber to catch the serrated edge of the beard and jerk it out through, and drop it again so readily by the back stroke.

Another plan for removing the beards was to sew a breadth of cotton cloth by the edge to a small rope, long enough to reach the whole length of the bolt; this was hung up along side of the reel on the descending side, and in such a position, that it might drag lightly upon the bolt cloth as it revolved.

This is the easiest plan, and the one most generally used at the present time. It is injurious to the bolt cloth if kept continually dragging upon it, and as an occasional sweeping is found to be sufficient, it is fixed so that it can be drawn up or let down at pleasure. When first let down after being long withdrawn, the bolt should be run empty for a considerable time, and then carefully shaken down and swept out. From a want of this precaution we have seen many a good large grist spoiled.

To remove the pasted flour from these cloths, spirit of wine or turpentine was used, applied to the pasted portions plentifully with a sponge, and when perfectly dry, as it would be in a few minutes if the spirit was pure, the part was switched smartly and lightly with a small switch until it was clean. There was no danger of injuring the English or muslin cloth by the concentrated spirit, but it must be sparingly and carefully applied to the Dutch anker cloth, or it will dissolve and destroy the sizing which glazes the thread, and fastens the meshes of these cloths, and constitutes their superiority and the great secret of their manufacture.

The safest and best application we have ever used is clean cold water, the only draw back being the difficulty of its application. We have in some extreme cases taken the cloth off; this is the most effectual means of cleansing it, but it involves much time and trouble, and great care must be taken to keep it thoroughly stretched while drying, or it will be too small for the reel. The Dutch anker cloth being now almost exclusively used, and the silk threads composing it being glazed smooth, and cemented at the crossings to preserve the meshes intact, they are neither liable to get very foul nor to be disturbed by brushing, and all that is required to keep them in good working order is the occasional use of the cloth sweep referred to, such washing as that referred to being only rendered necessary by an accident or extreme negligence.

The case was very different when the other substitutes were used for bolting cloth, and the smutting and cleaning apparatus was equally defective; the clogging of the bolt cloth was then a frequent occurrence, and this made the miller expert at cleaning it; but millers, like other craftsmen, are reticent of the particular sleight of hand they may acquire individually in such emergencies, and do not make it public. We mention these things because we have seen modifications of all these devices offered as new discoveries, and certain sums of money demanded for the receipts and right to use them, by speculators who had got hold of the ideas by some means, and sought to make money of them. But we have said enough on these introductory subjects, and will now proceed to give a description of the bolt as it is now made.

The reel is from twelve to eighteen feet long, and from two and a half to three feet in diameter; the shaft from four to six inches in diameter, is made of some light stiff wood, either six sided or round, with iron journals in each end, and the ends banded with iron; three or four sets of arms are mortised through this, one set near each end, and the other dividing the distance between these; each set is not placed around on the same line like the spokes in a wheel, but a single mortise running through is only made at one place, and a piece the whole diameter of the reel run through this to the centre makes two arms; another of the same kind on each side of this completes the set; they are put through in this way to be strong and light, and to avoid weakening the shaft. The ends of these six rows of arms are tenoned, and ribs the same length as the shaft mortised on to them. An end composed of six pieces of half inch boards is put around the head end; these are halved together at the ends for additional strength, and straight from rib to rib around the outside, but circled out round in the middle to admit the grist spout; this is fastened upon the end of the ribs, and six other pieces about four inches wide are nailed around the outer edge of these, each piece reaching from one rib to the next one, the ends being fitted and fastened to these. This makes a strong light corner on which to tack the head end of the bolt cloth; the tail end requires only straight strips nailed on from rib to rib to fasten the cloth, and leave the inside clear for the bran to pass out.

There are several ways to put the cloth on to the reel;

sometimes the cloth is split lengthways in strips a little narrower than the distance between two ribs, a piece of strong double cotton is sewed in between each of these to correspond with the ribs, and through which the tacks are driven; the last two edges have a strip of the same sewed upon each, and along the edges of these a row of eyelet lace holes are wrought to lace it together. A piece of the same cotton is sewed around the head end wide enough for the grist spout to discharge upon, and a narrow piece around the tail end.

To put it upon the Reel.

Take the cloth and fold or roll it up to the middle, beginning at both ends; take this roll and fasten the end to a rib near the middle of the reel, and turn the reel round, keeping hold of the roll until the first end comes round within reach; draw a lace through a couple of the eyelet holes about the middle of the cloth, and fasten the two ends of the roll together, and turn that splice side down. Now take hold of the head end of the cloth at the top of the reel, and draw it along to the head end and fasten it with a few tacks; stretch out the other end to the tail of the reel the same way, and fasten it slightly; now draw the edges together by the eyelets, a loop here and there will answer, and begin at the centre of the head end to fasten permanently around the end rim each way, drawing the laces near the end to bring the cloth to the proper tension before the tacks are driven. When the head is all securely nailed, send a man to the tail end to draw the tacks holding that end, and have him pull upon the cloth at the centre rib while you commence at the head upon that rib and rub hard with the palm of the hand along the reel toward the tail end, repeating the rubbing and continuing the pulling until all the wrinkles are out, and the cloth is tight enough; then that spot may be fastened, and all the other ribs served the same way until that end is fastened all round like the other. The two sides may now be tightly laced and drawn together, to stretch it sufficiently the other way, and the whole nailed to every rib, the whole length, with six ounce tacks, three or four inches apart. The ribs should always be made of soft wood, because the spring of the unsupported portions between the arms is so great that it is difficult to drive the tacks into hard wood; besides, it is almost impossible to get the tacks out of hard wood without tearing and injuring the cloth, when it has to be taken off.

This method of putting on the cloth requires the reel to be accurately made, and the measurements very correctly taken, and very carefully worked out upon the cloth, in order to have it all come out right; in fact it requires a good seamstress to sew all the different strips together without gathering or puckering, even when they are all cut out right, and a greenhorn would make bungling work of it.

When a reliable hand was not to be had to prepare the cloth in this way, we have frequently put it on as follows: Take a tape or other line that will not stretch, and measure round the circumference of the reel, and cut the cloth in lengths half an inch over the length of this line; have all these sewed together side and side in the order, with respect to the number and quality of cloth, in which they are to be placed upon the reel, and have a wide strip of cotton, as in the other plan, sewed on the head (fine) end, and a narrower piece on the tail end. Place this upon the reel like the other, and take a large needle and good double thread and draw the edges together over the last rib, folding or rolling the two edges together and sewing "over and over" as it is called, until it is drawn sufficiently tight. Stretch a piece of tape, or strip of double cotton along each rib and around both ends, and nail through these, and the job will be finished. This is more easily done, and will last fully as long, probably longer, as the cloth always gives out next the seam of the double cloth first, on account of the accumulation of dust and worms, and other vermin which invariably breed here. We used formerly to paste strips of cotton along all the ribs before we put the cloth on, but found these to be a damage, for the above reason.

With respect to the selection of the numbers or qualities of cloth most suitable, the quality or proportion of each kind, and the most proper distribution of these numbers upon the reel, much diversity of opinion exists among millwrights and millers. Some place very fine numbers next the head of the bolt, and coarser and still coarser as they approach the tail end. Others will place three or four medium Nos. (eight or nine) all together, and all alike, reaching from the head as far down as the flour is fit to mix, or use, and then a No. seven and a No. five to sift out the coarse flour, and a narrow strip of No. 0 to separate the shorts. Another plan is to place one width or two of No. 8 next the head, and two or three widths of No. 9 adjoining this and reaching as far as the flour is fit to mix, filling out the tail end like the others.

In order to determine the best selection and arrangement of cloth for any particular case, we must consider the kind of flour principally to be made, whether one or more qualities, and also take into account the following principles and peculiarities in the process of bolting:—

When the grist is run into the bolt it is a mixture of

particles of very different degrees of fineness, and these are quite as different in their degrees of weight; this difference of weight is made to perform an important part in the separation of these particles, by gravity; the difference in size completes the separation, mechanically, by the intervention of the different sized meshes of the bolt cloth. When the mixture falls upon the bolt cloth, the fine flour, being the heaviest, strikes it first, and the particles of coarser flour, shorts, and bran, being lighter and larger, strike on top of the flour, and as they slide round on the inside of the cloth as it revolves, the different qualities maintain the same position with respect to each other, the finest being always next the cloth. Thus the finest of the flour will always sift through first, whatever the texture of the cloth may be, and the coarser, with specks of shorts and bran in it, cannot get through until the fine which intervenes between the cloth and it, is out of the way; but now the coarser flour and shorts falling and sliding in actual contact with the cloth, a grade of flour will pass through too coarse and full of specks to be mixed with the rest, unless the cloth be very fine, and hence the propriety and policy of interposing a finer breadth of cloth at this point. We have sometimes added from one to two pounds to the average yield of good flour from a bushel, by taking out a breadth of cloth here, and introducing one a number or two finer, and oftener made the necessary improvement by exchanging the coarser number here for the finer one next the head. Interested parties are slow to comprehend this subject, and persist in placing the finest cloth where the finest flour will come, without testing the matter either by experience or philosophy.

We may mention here another contingency with respect to bolts, although it might rather be mentioned in connection with the machinery for driving them. It is this, the bolt machinery should be so connected with that driving the stone, that when the stone is stopped the bolt will also be stopped, or if the stone has to grind other kinds of grain not requiring to be bolted, then some provision should be made for throwing the bolt out of gear. If this precaution be neglected, and the bolt run after the grist is done feeding in, the contents will all be sifted through, and the remaining light stuff, having no flour between it and the bolt cloth, will be partly sifted through, covering the flour with specks and spoiling its appearance and quality. The same thing will occur with the first entrance of the next grist run through it, a part of the light stuff will drift along the naked cloth, in advance of the main stream, and pass through the cloth in specks like the other; this ending and beginning will affect on an average half a bushel of flour, and where it often occurs it becomes a serious item. From these considerations, and a pretty extensive experience, we would recommend the distribution of the cloth as last indicated, for an ordinary custom mill, to be modified of course by circumstances, such as the quantity to be bolted in a given time, the kind of wheat or grain chiefly to be ground, &c.

We might give as an example of a bolt for an ordinary run of stones, say eighteen feet long, one width of No. 8, three of No. 9, half a breadth of No. 7, half a breadth of No. 5, and half a width of something coarser to let through the shorts. This would bolt faster to put on two widths of No. 8 and two of No. 9; the quality will also be finer the faster it has to bolt, and of course the flour will be coarser the smaller the quantity passing through it. The like alteration of coarser and finer can also be made from the same bolt, by elevating or depressing the tail end, thus causing the contents to pass faster or more slowly through. The quantity of each kind of cloth is easily computed; it will average three feet three and a half inches in width when sewed together, and if the reel be thirty inches in diameter, it will take two and a half yards to go round it; if thirty-three inches, two and three-quarter yards; and if thirty-six inches, three yards.

The bolt should make from twenty-eight to thirty-three revolutions per minute, according to its diameter; but there is more damage done by running it too fast than too slow, because the centrifugal force prevents the contents from falling and sliding upon the cloth; the proper motion allows the stuff carried up by the ribs to fall down again about the centre of the bottom; if much is thrown on to, or against the shaft, the motion is too rapid. The best method of gearing a bolt is to take the motion from the spindle of the stones connected with it; to do this, place an upright shaft at the head end of the bolt with a large and light belt wheel two and a half or three feet in diameter upon the lower end, place a small sheave to correspond with this upon the spindle; this is best done by turning a pulley the right size, marking the size of the spindle upon both ends, and also a seat for a narrow band upon each end, then split it into two pieces, and dig out the centre to the spindle marks, then put the two halves together on the spindle, and put on the two bands (previously fitted) one above and the other below the pulley, rivet the ends together upon the spindle, and drive them on to their places; place a bevel pinion on the upper end of this shaft, and another gearing into it upon the shaft that carries the elevators. Place a light spur wheel upon this shaft, and a corresponding one upon the journal of the bolt shaft;

these two may be geared together, but it will give more room for the elevator spout, and bring the motion of the reel in a better direction, to interpose another wheel. The intermediate wheel may be made to slip endwise out of gear with both the others; or it may be permanently geared into the elevator wheel, and the end of the bolt shaft with its wheel made to raise and lower, into, and out of gear. This is done by hinging the back end of the supporting bar on a pin, and raising or dropping the other end by a handle fixed for that purpose, slipping a key under, when it is up and in gear, and the same key over the bar when it is down.

By this arrangement the same stones and elevators may be used to grind Indian meal or feed, by having a crotched spout at the discharge of the elevators with a swing gate or trap valve to throw the grist either way, into the bolt or the meal box. The bolt is frequently driven from the elevator shaft by a belt, but this is more troublesome, and not so reliable, besides the slight jarring of the cog gearing is an advantage, and tends to keep the cloth clean. The elevator shaft should turn faster than the bolt reel, otherwise the upper elevator pulley must be inconveniently large to give the cups the requisite speed; both pulleys should be a little larger than the space between the two interior sides of the cases, to clear the belt of the corners; the belt should also be a little narrower than the cases, and the cups narrower than the belt. An elevator should always, if possible, be driven by the upper pulley.

The most common material used in making the cups is tin, this is the best for grain, and in ordinary circumstances for flour and meal also; but when damp grain is ground rapidly, and elevated through a cold atmosphere, a steam is generated which condenses upon the cups and causes the dust to adhere, forming a paste which accumulates and fills them up, and is a source of considerable annoyance; in such a case leather is preferable to tin. When well made of good leather, they answer the purpose well, but we have often seen them of such a make, and of such material, that they were a great nuisance. For such situations, the cups are frequently made of wood. To make these, take a strip of bass-wood, buttonwood, or poplar, two inches thick, and the width of the required cups; bevel off the corner something like the shape of the ordinary cups, and saw it into lengths; a little under the breadth of the belt; hollow them out slightly above and behind in imitation of the tin cups, and fasten them by small nails screwed or driven through the belt. These are not liable to clog up nor catch in the cases, and may be placed on the belt close enough to make them carry as fast as required.

The bolt chest may next be considered. It incloses the bolting reel or cylinder, and confines and collects the flour as it sifts through the bolt cloth into the interior of the chest, from whence it is shaken out into the flour through in front, by drawing up a range of slides. In this trough the flour arrives exactly in the same order that it passes through the cloth, the finest next the head, and the grade becoming gradually coarser as it approaches the tail end where the bran and shorts are discharged. This enables the miller and owner (customer) both to examine and select the most suitable point to divide the flour, or determine how much coarse should be taken off to suit the customer's taste and economy. This is a great advantage in a custom mill, where almost every grist differs in some respect, and where such a diversity of opinions is held by the different owners.

One is a poor man, or what amounts to the same thing here, a tight man, he wants every pound of flour mixed in that is fit to use; the next one is not particular about the quantity of flour, but wants it of the very best quality; here the miller can suit both these customers, and have their own judgment and sanction to back his own, which is a good recompense for the extra trouble of mixing and filling the flour by hand. To shirk this labor, and have the flour mixed and filled into the bags by a conveyor, the miller must forego the satisfaction of pleasing his customers, and suffer the positive annoyance of not being able to suit himself instead of suiting his customer in many instances. This is more particularly felt in grinding small grists, when he may examine the flour and set the slides differently, and find out about the time the grist is done how he might have done better, but instead of profiting by the discovery, he must now put in another grist of a different quality, and for a different customer, and go over all these manipulations again, probably with a similar result.

The screw conveyor bolt, in addition to this uncertainty as to the proper point to divide the flour, has another objection, which is that it mixes two succeeding grists together, so that when the grists are small it is impossible to give a customer the product of his own grain. This would be of little consequence if the grain were alike, but when one is clean perfect wheat, and the other a mixture, of comparatively little value, then it is a positive injustice: because the man who has a small grist of good wheat ground after that of inferior quality, gets inferior flour, while the next, with a small grist of poor stuff, perhaps will get his good flour.

For this reason, the conveyor bolt, which mixes the flour so perfectly, and saves so much labor in filling when large grists of a similar quality are manufactured, should never be used for a custom mill, where small grists of different qualities are often ground.

We are anxious to place this subject plainly before parties intending to build, for the reason that a kind of mania for this kind of bolt passed through some sections of the country many years ago. These have nearly, but not all, been superseded by or altered to the trough bolt; and in making these alterations we have had our full share of trouble. The mania appears lately to have broken out again in other sections, and we would recommend parties interested to consider the peculiarities and contingencies, which are here briefly pointed out, before deciding which kind will answer their particular purpose best.

The medium size and proportion for an ordinary trough bolt chest, is about eighteen feet long, three and a half wide, and nine feet high, the bottom of the trough about two feet from the floor, and the upper commencement of the angle for the slides about two feet above the bottom of the trough; the trough is about fourteen inches wide at the bottom, and twelve deep, the front flaring out three inches further at the top than the bottom. The front is sometimes made of panel work of two inch plank filled out with inch boards; a more common, and probably the best way, being to make the framework of scantling, the front and ends boarded on the outside, and the back on the inside of these frames. The slanting portion which gathers the flour forward to the trough should commence pretty high on the back, and terminate on a perpendicular rise of three or four inches at the back side of the trough. This is necessary to get the incline steep enough to cause the flour to run freely down into the trough, and the perpendicular rise at the back is an advantage in mixing and shovelling out the flour.

It would be difficult to describe the exact form and construction of the framework of the chest, but it is not requisite, it being generally understood by practical workmen. A range of doors should be made along the front, opposite the reel, and corresponding with the flour slides below. These are required to get at the reel, and should be large enough to allow a person to work with freedom. A light neat covering for them is a frame made like a picture frame, and covered on the inside with pink or fancy colored cambric, which should be hung by hinges at the upper edge, or otherwise fastened by a neat fancy button at each end. Such doors are preferable to any that can be made of wood, because they tend to cool the flour by the ventilation they induce, and at the same time sufficiently confine the dust. The top of the bolt chest should project over the front a little further than the trough, otherwise the worms, insects, and mice working upon the top will push over dust and dirt into the flour or empty trough. This projection may be carried round the ends, and a tasteful moulding put around it like a cornice, costs but little and improves the appearance.

Another very necessary appendage of the bolt chest, and one that is often neglected, is the "Speck box." When the grist enters through the end of the reel, it is carried up by the revolving ribs continually, and falls down again past the opening in the end, and a portion of the fine light stuff dusts out through the opening. Among this dust there are always thin light flakes of bran and shorts, which show themselves among the flour by the contrast of color. The ordinary method of boarding up the end of the bolt chest on the outside of the frame, gives a good chance to make a little box or division to intercept these, by boarding up on the inside of the end frame, and rounding out and fitting the upper edge to the circle of the reel. This speck box must have an outlet at the lower point, outside, and a small spout to convey the contents down into the lower or back side of the elevators. Although the utility of this speck box is so obvious, and its construction so simple, many first-class millwrights neglect to put it into custom mills, and leave these drift specks to mix with the flour.

Before leaving this subject, we might mention a few of the accidents by which the bolt cloth is most liable to be damaged or destroyed, as it is a very fragile fabric, as well as very expensive, and should be very carefully protected. The most common accident to which it is subjected is by some heavy sharp instrument, such as a chisel or file, being dropped into the spout, elevators, or on to the running stone, and carried through the bolt. We once saw a bolt cloth badly cut by a sharp-pointed jack-knife which the miller dropped upon the stone; he supposed that it lodged between the stone and curb, and did not stop the mill until he discovered by the bran among the flour, that it had passed through the bolt and cut a number of small holes in it. In another instance a small newly ground mill-pick passed through and mangled the cloth badly; many of the blemishes which this caused were so small that they escaped detection for some time, and after frequent examinations.

We have seen several bolt cloths pasted and totally clogged up, by standing idle and partly open in cold weather, and having a door or hatchway communicating with the water spray carelessly left open; the steam or dampness in the atmosphere condensed upon the cloth, in small needles, tangled together and forming a kind of frost rhime. This could all have been absorbed and removed by circulating cool dry air through the bolt and apartment, leaving the cloth unharmed. But the injury was done by not knowing, or not minding this, and thawing it out suddenly so that the moisture completely saturated the cloth; the dust adhering to it rendering the whole as impervious as pasteboard. These were the cases referred to, where the cloth had to be taken off the reel and washed out with cold water. Only one case ever occurred to our knowledge of a bolt cloth being destroyed by fire; that was at the time when muslin was used for bolt cloth, and is not likely to occur with Dutch cloth, which burns like woollen or feathers. The circumstance referred to took place with a brother of ours. We were attending a mill for our father, and being only boys we had to assist each night alternately; my brother was grinding a grist of very dry buckwheat, and running it through a muslin bolt kept for coarse grain; some black specks in the flour showed a defect in the cloth, and he opened one of the doors to ascertain the cause, while the bolt was still in motion; as he raised the candle or lamp up near the cloth the dust took fire like a flash of powder or gas, and both dust and bolt vanished in an instant, in a cloud of smoke, leaving not a vestige of the cloth, and only the naked skeleton of the reel.

We have never seen or heard of a similar accident since. But anything that has once occurred may occur again under similar circumstances, and for that reason it may be worth recording.

Another important apparatus in the grist-mill is that used for cleaning and scouring the grain preparatory to the grinding. There is an almost endless variety of these machines, constructed on many different principles, and known under the general name of smut machines and separators. Many of these are patented, and some are very excellent machines, leaving little more to be desired in the way of improvement, and in a measure relieving the millwright of another important branch of the business which formerly devolved upon him; for the making of these machines is also a separate branch of the business, carried on in distinct factories.

For this reason it will not be necessary to give a detailed description of the origin of these, nor to trace them through the various modifications they have undergone within the last forty years in their progress towards their present degree of perfection. We have within that period *made* not less than ten or twelve different kinds of smutters, some of them running horizontally, but more vertically, and have *seen* at least ten times as many varieties made by other parties. A detailed description of the most important of these might be interesting to some readers, but as they are mostly superseded and out of use, we will not attempt to rescue them from oblivion, as the benefit could hardly be sufficient to repay the time and space thus occupied.

We will, therefore, only briefly describe the process undergone by the grain for its purification, and the principle upon which these machines are constructed. The first operation is that of screening. To effect this, the grain is passed through a coarse screen, the openings of which are only large enough to admit the passage of the grain, and exclude everything else larger, which must pass *over* instead of *through*, and is thus separated. It is next passed over a finer screen, the openings of which will not admit a kernel of grain, which has therefore to pass over it, while everything smaller than the grain, drops through and is also separated; by this part of the process everything larger and smaller than the particles of grain is separated. But the "smut" proper is the same size and shape as the grain, and cannot be screened out, and therefore the principal business of the smut machine is to demolish and separate these "balls" or kernels of smut. These are covered by a husk or hull, similar to the grain, but the inside is filled with a soft black powder resembling snuff, and it is the softness and friability of the interior of these balls that is taken advantage of to effect their separation from the grain.

The machine consists of a cylinder which is made to revolve inside of a stationary case; the grain is admitted into the space between these at one end (the top), and works its way through to the other end where it is discharged. Both case and cylinder are furnished with angles for the grain to strike against, and the velocity must be sufficient to break the balls of smut against these angles, but not sufficient to break the grain, which would thus be wasted; the breaking of these balls liberates the smut, which must be instantly separated by a blast of wind, or otherwise it will be daubed on, and adhere to the grain, from which it can only be removed by a mixture with lime or other similar absorbent; and on the speed and perfection with which this separation is effected, the main difference of the various machines depends.

A scruffy outside portion of the bran, particularly at the blow end of the grain, is also detached by the battering which it undergoes, and this, together with the broken and imperfect grains, passes off in the blast of wind with the smut. Hence another criterion by which to decide the relative merits of the various machines, is the degree of perfection with which they separate and save the useful portions carried out by this blast, and confine and dispose of the smut and useless matters which when blown about the mill are a great nuisance.

It is impracticable to describe *all* the good smut-mills which are for sale, and to single out one or two would subject us to a charge of partiality which we do not care to incur, particularly as this branch of the business is in a manner removed from our particular province, and is established as a separate branch of manufacture. We prefer therefore to leave parties to make their own selection, having stated the principle upon which they work, and indicated the essential points to be regarded in comparing and choosing. All that is necessary will be to get a pamphlet with descriptions of good machines, etc., which will enable each inventor to show up the peculiar merits of his own machine.

The screen is not so easily disposed of; it has not been improved like the machine itself, and several that were sent along with some of the best and most expensive machines that we ever set up were not used at all; nor would they have been used by any good millwright forty years ago. These were shaking screens to be driven by a crank; they were three feet long and one foot in width, the upper (coarse) one of zinc punched, the lower one of wire cloth. The material and make of these were well enough except that the upper one was too coarse to do any good, and the lower one so small, that one grist mixed with small seeds or shrunken grain, passing over it would fill every mesh in the wire, and after that the grain might as well be screened over a board. Had these contained thirty-six superficial feet each, instead of three, then the upper one could be punched fine enough to screen everything out that was larger than the grain, and yet allow the grain to pass through fast enough, and the lower one would not get altogether choked up until the miller would get time to brush it out, by rubbing the under side with an old broom or piece of wood.

The shaking screen is much more liable to clog up in this way than the rolling one, and for that reason it should always be placed so as to admit of being conveniently brushed out. Zinc or tin evenly punched, without burr or bulging, is the best material, particularly for the upper coarse screen. This, with a proper motion and declivity, enables nails, sticks, or straws, or even oats, to sail along and pass over, which would drop endwise through the same size of holes in a rolling or revolving screen. There are many ways of hanging these, but the simplest and best is to hang them on the top of four wooden springs. These should be made with a true taper the whole length, the butt (thick) end mortised into something substantial on the floor, and the screen poised upon the thin ends above. This may be driven easily and steadily by a single crank attached to it by a connecting rod (pitman), while many of the other modes require a double crank, that is, a crank on both ends of the same shaft, and two connecting rods to insure a steady motion. Sheets of zinc or tin can be bought already punched by machinery, or if not, they may be punched by hand, with a steel punch upon a large cake of lead, or upon the end of a large block of hard wood. The wood must be cut off occasionally when it gets indented too much by the punch, and the lead must be hammered smooth.

Merchant, or Manufacturing Bolts.

Although the principle by which the Merchant bolt separates the different grades of flour from the bran and other offal is the same as that by which the little custom mill bolt effects the same purpose, yet the construction and details of the one differ greatly from the other. The constant and rapid rate at which the first-named works, and the perfection with which it is required that each ingredient must be separated, and conveniently disposed of, make it necessary to employ an extensive combination of complicated machinery, quite different from the simple single reel and chest which suffice in a custom mill.

The vast competition in milling makes it necessary that the machinery by which the flour is made be such as will yield the greatest quantity of the best quality of flour from the wheat, and that in the shortest time and with the least expense in the operation.

Recently a great many improvements have been proposed and tried in bolting; some of them are patented, but most are improvements only under certain circumstances, while some are worthless. One mill man will approve of one of these, that another of equal experience will condemn. This is owing to the different conditions of the mill in which they are tried; one patent is for inside knockers on a bolt reel: small iron weights to slide on iron rods, fastened one end in the shaft and the other end in the rib; as the reels revolve, the weights slide down the rod from the shaft and strike the rib inside, jarring the flour through the cloth; this is of use where there is too little bolting capacity, and it fails to clean the offal. Perhaps the miller in one establishment has allowed the stones to get smooth, the agent putting in the improvement will see that they are well dressed and balanced, and will then start the mill with plenty of bolting room; while another miller, having plenty of bolting capacity, and the stones in good order, might apply these knockers, and they would spoil the flour, by jarring through the specks.

Where a mill has too much bolting surface and makes the flour specky, it may be cured by putting the cloth on the inside of one rib, and the outside of another, so that the flour will slip over each alternate rib without being lifted; the flour will be cleaner, but it will not bolt so fast. Sometimes the cloth is put around inside of all the ribs; of course this doubles the effect of the other plan in the same direction.

The process of bolting seems to admit of more variation than any other branch of milling; the speed, for instance, may be varied from eighteen or twenty revolutions per minute, to thirty-four or thirty-six revolutions, without apparently making much difference in the work. Close observation might detect an error in such wide variations, and they would be inadmissible in merchant bolts where the load and quantity ground are kept almost constantly the same. It is believed by many good millers that the merchant bolt has not been brought to the perfection it might be, and that some ingenious mill man will yet perfect it and make a fortune by it; our opinion is, that he who attempts it had better try to devise some entirely new system, as there is a danger of the old getting too complicated if it be much further *improved.*

The object to be kept in view in constructing these bolts, is to make a complete separation of the meal at one operation, dusting the middlings at the same time. After middlings are separated from the flour and bran, a good deal of fine flour, which ought not to be ground over, can be sifted out by passing over fine cloth; the same is the case with the bran. This fine dust cannot be separated so easily while the bran and shorts are together, because a portion of the dust as it is detached from the middlings adheres to the bran, and it would have to pass over a great amount of cloth in that situation to separate it; this holds good with middlings and flour, and the sooner they are partially separated, the better. Separating the bran before bolting has been tried; we have had no experience with this plan, but think it will eventually be adopted to a certain extent.

In every merchant mill, some kind of apparatus must be made to intervene between the grinding and bolting, to cool the meal. There are some *extra* millers and millwrights who say that the best way to cool the flour that can ever be devised, is *not to heat it;* that is impracticable, if not impossible, and we must either select one of the several ways now in use, or else provide a better. Many of the large merchant mills still use the old-fashioned cooler, similar to that previously described in the chapter on grist-mills, the only alterations we notice being unimportant, as the cams for shaking the feeding shoe, and sometimes a new-fashioned rotary device for that purpose. The cooler answers as a reservoir for the meal, when the miller does not grind as fast as he wishes to bolt, which is often the case. Some use lines of open conveyors for coolers; the principle is as good, or perhaps better, than the old way, but lacks the reservoir, which might be provided with a proper garner above the bolts.

A better plan is, a blast of air, similar to the fan elevator, only do not attempt to raise the flour through three or four stories, and perhaps blow it as far horizontally, but raise it up through one story by the blower, and carry it the remainder of the distance by elevators. This will allow the fan to run at a moderate speed, and cool the meal perfectly, besides drying it, which is also very necessary, especially at the West, where grain is not housed before threshing, and often indifferently housed afterwards; the consequence is that the flour, if barrelled immediately and shipped, is liable to sour.

This drying process helps to whiten the flour, and passing through the fan at a very rapid motion, scours the bran; the only objection is, that in cold weather the steam condenses on the inside of the tin pipe and clogs it up; this can be prevented by covering the pipe with some non-conducting material. Some provision should always be made for regulating a cooler to the amount required, as too much stirring and cooling in very dry or cold weather make the meal bolt too freely, and the flour specky; this can be tempered in the fan cooler by taking the air at times from the curb, or from the outside atmosphere. In concluding these remarks, we may say that the principle of the fan cooler is good, but the details must be carefully managed; and finally we would advise the avoidance of heating as much as possible, by keeping the stones sharp, true in face, and well hung, for no matter how the flour is stirred and cooled, if it is not properly ground it will not bolt well.

We may here mention a simple and useful device for carrying the steam and dust away from the curbs and conveyor in front of the stones, by which means the mill apartment and everything it contains are easily kept clean. A suction fan is placed in the upper part of the building, with a pipe leading from it to the top of the conveyor in front of the stones; this pipe passes perpendicularly from the conveyor through the floor above, where it is discharged into an air-tight room; another pipe is taken from another part of this room to another similar room, and from thence to the fan, and thence out through the building where it discharges. The air-tight rooms retain and deposit all the dust and flour that are carried from the conveyor. This arrangement also tends to modify the heat generated by the friction and pressure of grinding, as it increases the circulation of air within the curb and around the stones, and this, with its cleanliness, will insure its general use in well finished mills.

There are several variations in the structure of these bolts, some having whole chests containing five or six reels, arranged and so connected that each performs its part in unison with the others, and the process is completed at a single operation. In others the reels are divided into half chests, or otherwise, the operation being divided in like proportion. A five reel full chest is frequently made with four reels, thirty inches in diameter, and twenty feet long, the two upper reels covered with No. 10** cloth, the entire length; the next two with No. 12** cloth, the entire length; the middlings and bran falling together from the tail of these last reels, into a separating reel and duster. This last is forty-four inches in diameter, the head covered with No. 10 cloth, and the tail end, where the middlings are separated from the bran, with No. 5.

We give the plan of a six-reel whole chest merchant bolt; it is similar in most respects to that just completed in the River Street mill, Milwaukee, Wis., by Henry Smith, Jr. The reels are forty-four inches in diameter, and twenty feet long, and being all in one chest, make it very high, and it extends up into the next story of the mill; the two upper reels are used as flour reels, taking nearly the entire length, covered with Nos. 10 and 11 cloth. It will be seen by the drawings that what falls over the tail end of the upper reels drops into the return reels directly under the others; two-thirds of these are covered with fine cloth, the balance with No. 5, which separates the middlings from the bran. The middlings are carried to the tail end of the return reels, both joined in one spout, and thrown into *one* of the two remaining reels, called the middlings dusting reel, which is covered with very fine cloth; the bran falls over the tail end of the centre or return reels into one spout, and is carried into the *remaining reel*, which is also covered with very fine cloth, and is called the bran dusting reel, this makes the full complement of six reels.

The return business is so managed that all the stuff coming through the return reels *proper*, and what is dusted in the remaining two reels, is carried to one spout on the bolt floor, discharged thence into the conveyor in front of the stones, and carried back into the cooler. This arrangement is capable of working over five hundred barrels per day, and is the handiest merchant bolt that we have seen.

The greatest trouble in all large mills appears to be the working up of middlings. In good milling times owners can afford to make one first-class grade of flour by grinding high and taking only the head of the bolts, and then making two or three other grades of flour out of the middlings; but when flour is cheap, and there is little demand for inferior flour (or in fact at any time), the object is to get all the first quality flour out of the wheat without injuring its color.

The following arrangement is the best we could devise for this purpose: Presuming a mill has seven run of stones, five for wheat and two for middlings, it would require one full chest of bolts similar to Smith's plan for the wheat stones, and a half chest of bolts for each of the middling runs, and three coolers. One run of middling stones should grind up the middlings as they come from the first bolts, just fast enough to keep up, and no more; these would yield a grade of flour clear enough to mix with the first flour without rebolting. There would still be another grade of middlings left, and another run of stones and half chest of bolts left to work these up.

A separate grade of flour may be made out of these last middlings, or it may be run into the first and keep a XX grade up, by bolting in this half chest and running the flour back into the cooler, and rebolting with the meal from the wheat stones. Thus the whole can be run into one grade and still ground high enough not to injure the color of the flour.

The attempts made to grind close and soft enough the first time, bolt close, and only grind the middlings once, and clean them, running them all into one grade, have generally failed, and cannot be relied upon, being successful only when all the circumstances are favorable.

We intended to insert a plan and description of a half chest bolt, which answers well in mills doing both merchant and custom work, and also some further information on cooling and packing machinery; but millwrights superintending the erection of such mills are generally quite competent, and frequently have their favorite systems—to such, this chapter will be of little value, and will be more useful to young men, and those of limited experience; for these reasons we will add no more here.

CHAPTER XVII.

THE OATMEAL MILL.

THE process of manufacturing oatmeal is but little known or practised in these United States, although it is carried on to some extent in the adjoining Dominion of Canada. The first essential part of the operation is to divest the grain of its outside hull, and clean the kernel of the stratum of dusty down in which it is enveloped. When this is accomplished, it is ground and sifted like Indian meal, only much faster and with less power.

The first preparation requisite, is to expel the moisture from the grain, until the kernel is hard, and the hull stiff and rigid. In this condition the hulls are easily knocked off in passing through the shelling stones, and the down referred to, is displaced and reduced to dust by the attrition undergone during the operation. As this mixed mess leaves the shelling stones it is passed over a shaking screen of fine wire, that separates the dust and any small pieces that may be broken from the grain; it then passes through a winnowing machine, similar to an ordinary fanning mill, which blows out the hulls, and delivers the grain clean and ready to grind into meal.

Although the grinding of oatmeal requires *less* power than Indian or other meal, it also requires *more* attention and care. As it is dried and divested of its covering, it is easily ground to powder, but for excellence and long keeping it requires to be ground coarse and round; and for these reasons the best oatmeal is ground either in shelling stones, or in stones dressed and hung similarly to them. That is, the stones are dressed as sharp as possible, with a little *bosom*, but otherwise true and equal on the face, without any furrows, and hung upon a stiff ryne instead of a balance.

When ground, the meal is sifted through several sieves of sheet metal, punched with round holes. These are placed one above another, in the same frame, which is hung by four slings from above, and a rotary swing given to it by a short stroke crank, the axis of which is perpendicular and above the centre of the sieves.

The Kiln.

The kiln is the most important and expensive apparatus required in the process of manufacturing oatmeal. It is built of stone or brick, from sixteen to twenty-four feet square, and about twelve feet high, without any other opening than the furnace door, which is generally walled back in a recess about six feet toward the centre. Here the furnace proper commences and runs back to the centre, where it is extended up and spread out into a *lantern*. This lantern, like the furnace, is built of brick, and provided with small flues, which scatter and distribute the heat equally throughout the interior. The beams and joisting to support the floor are of bar iron, supported at intervals on iron posts, and the floor is either composed of large square tiles or sheet iron, in old kilns, or cast-iron plates in the more modern. Of whatever material the floor is composed, it must be thickly perforated with funnel-shaped holes, the wide end down, to allow the heat and smoke to pass up and prevent the oats and dust from passing through or choking the holes.

The walls are generally built about three feet above the kiln head or floor, and a pavilion roof, like a hopper reversed, covers the whole, from these walls to near the centre, where an opening about four feet square is left for ventilation; this ventilator is roofed over some distance above, and its four sides are inclosed with lattice-work to exclude the rain and snow. A small ventilator should also be made over the centre of each wall (the door answering for one), the other three should be provided with a shutter hung by a horizontal axis through the centre; these will be shut or opened by the kiln-man to admit air and light, according to the direction of the wind.

In Canada, these kilns are generally heated by fires of wood, but some of the best millers have introduced the old country system of drying with the shelling seeds, or hulls of the oats. These being previously kiln-dried, produce little smoke, and no steam, which is a great nuisance when drying with wood. The objection to this plan is that it requires the constant attendance of a person to scatter in the seeds, and stir up the fire, but American ingenuity would soon obviate this objection by delivering this fuel through a tube, by a blast of air or otherwise, and thus save the expense of fire wood, and at the same time use up the oat hulls, which, when allowed to accumulate, soon become a great annoyance.*

Process of Drying.

When the kiln head is hot enough to hiss when water is sprinkled upon it (most kiln-men spit upon it for this test) the oats are spread on to the depth of five or six

* Under a proper and economical system of agriculture, these, if not used for fuel, should go on to the compost heap to form manure, and thus return to the soil its constituents.

inches, making from fifty to two hundred bushels, according to the size of the kiln, and the depth of the batch; the fire is kept up, and the moisture expelled from the oats next to the iron, condenses in the upper and colder stratum; this is called *sweating*, and after it has advanced considerably, the kiln-man, to expedite the process and equalize the heat, turns the whole batch over by tossing it up, a little at each flirt of his wooden shovel. Each batch is turned in this way several times, which helps to dissipate the moisture and prevent the lower strata from being scorched.

When sufficiently dried, which is soon learned by practice, the batch is removed and immediately replaced by another. The time varies according to circumstances, three batches per working day being about the average; the oats require time to cool before they are fit to shell.

The stones best adapted for shelling are a coarse, free, and moderately soft sandstone, those from the Newcastle quarries, in England, having the best reputation; although there is no doubt that stones of an equally good quality may be found in this country, while all that we have thus far seen from American quarries are too hard, and have a tendency to glaze.

Dressing and Hanging the Stones.

The stones used for shelling are about the same diameter as the ordinary burr mill-stones, but not so thick and heavy, as either shelling or grinding requires but little pressure, and any weight beyond that required for strength and safety is an incumbrance. The bed-stone is faced perfectly true, but the runner has a space or bosom at the eye, of about three-sixteenths of an inch running out to nothing at about two-thirds of its diameter, the outer third being ground true upon, and agreeing with, the face of the bed-stone. Both are picked rough with sharp, square-pointed picks, but have no furrows. The bed-stone is laid like a flouring stone, but the runner is placed upon a stiff ryne, with three, or sometimes four horns, the ryne keyed tight upon the spindle, and the horns let into open gains cut in the stone. (See Fig. 40.)

The stone and spindle must be carefully adjusted to the face of the bed-stone as follows: Turn the stone round slowly, and lower and move the spindle by the step under it until the lowest edge of the running stone touches the bed-stone equally all round; then the spindle is tram or true with the bed-stone, and the runner must be set perfectly true with both of these. To accomplish this conveniently, the runner should be drilled through over each horn of the ryne, and a screw bolt with a square head passed down through these holes, the nuts being fitted and fastened with lead in the stone above each horn, and the round point of each bolt resting on these horns. Raise the runner one-fourth of an inch clear of the bed-stone, and hold a thin piece of wood on the bed-stone between, then have the runner turned slowly round and mark the lowest or closest place, tighten the screw bolt over the horn next the low side, or slacken the opposite ones with a wrench, and turn and try, and shift again until the stone scrubs the stick equally all round, and it will be ready to run.

If no screws are put in to adjust the stone in this way, it is much more difficult to hang, as the low edge of the stone must be pried up, and pieces of tin or sheet iron, and lastly, slips of paper inserted between it and the ryne to bring it true.

In the process of shelling, the stones must be *the length of a kernel of oats apart.* The hulls are knocked off as the grain is canted over endways between the stones, in its whirling progress from the eye toward the skirt; the stones having no furrows to assist the outward

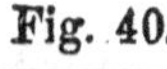
Fig. 40.

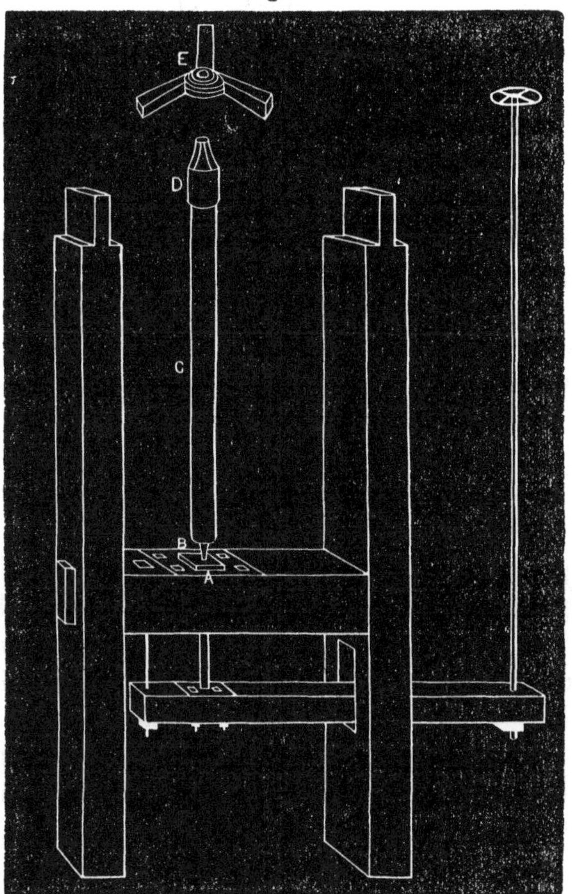

A, step; *B*, toe; *C*, spindle; *D*, collar; *E*, stiff ryne.

draught of the oats, they are impelled through by the centrifugal friction of the runner only, at the same time being retarded by the quiescent bed-stone, and each kernel further interrupted by those in advance of it and those tumbling over it from the rear. That this is the necessary condition of their progress through the stones, in the act of shelling, is proved by the fact that the shelling cannot be effected with an ordinary grinding feed, but the feed must be increased until the confused interruption indicated takes place, when every kernel has to turn one or more somersaults before it makes its exit, and thus the hulls are split off.

Many good millers run the oats through the shelling stones twice; the first time the stones are set high, the second time a little lower. By this means the oats are shelled cleaner, with less broken grains and waste. The duster and fan for blowing out the hulls are simple affairs, which may be made to answer the situation and circumstances, and are already sufficiently described.

It has been stated that the best oatmeal is generally ground in the shelling stones, or in stones dressed and hung in a similar manner; but if we were fitting up an oatmeal mill for our own use, our present experience would make us change the old practice slightly, for the following reason: The shelling is best performed when the face of the stones are as sharp and rough as possible, the friction upon and commotion among the grains as described above being thus increased; but for grinding, a smoother and finer face answers better, and it improves the shelling stone face for grinding, to run them slightly together, and thus face down the sharpest points before commencing to grind meal; and further, the meal is not benefited, like the shelling, by being retarded in its passage through the stones, but injured, as a greater proportion is reduced to flour, and for this reason the draught should be assisted by leading furrows.

We said the dried and shelled oats were easily ground; of course then the grinding is performed with little surface of stone, and this little surface is obtained in large stones by a large eye, and cutting out the bosom. We have seen a shelling stone with an eye twenty-two inches in diameter, almost large enough for the stones required to do the grinding; this stone was cut away at the eye one-quarter of an inch, and running out to a true face at two-thirds of the radius, the other third only being available for grinding and equalizing the meal; and to this point we wish to call particular attention. The grinding being performed between two narrow surfaces, these require to be perfectly true, and exactly the same distance apart at every point, as the stone revolves, in order to insure equality in the grist, and this is more easily attained in small stones and near the spindle, than in the skirt of large ones and far away from it.

To make this plainer, suppose the large stone be one hair's-breadth out of true at one-third its radius from the centre, it will be twice as much at two-thirds, and the breadth of three hairs astray at the skirt. Now it must be admitted that it is *possible* to tram and hang a large stone as true as a small one, but it is not possible to *keep* it so true when working. We seldom see a stone that has been run for any considerable length of time without the spindle requiring to be trammed anew, and the stone is equally liable to get out of true on the ryne or spindle; and when any of these things occur, the faces of both the runner and bed-stone get worn more on one side than the other. When all these contingencies are combined, it will be seen how difficult and next to impossible it is to keep these narrow ledges of face around the skirts of large stones sufficiently true and parallel with each other when working.

For these considerations then we would use separate stones for shelling and grinding, and instead of having the shelling stones from four to six feet in diameter, we would have them from three and a half to four feet, and those for grinding from twenty-eight inches to three feet, according to the quantity to be made, and other circumstances. Of course this diminished size of stone would involve a corresponding diminution of eye and bosom. The power required to work these two run, with the necessary cleaning and sifting apparatus, elevators, &c., would be about that required for one run of flour or corn stones.

We have heard of shelling stones made of wood and faced with a composition of emery, which are said to keep sharper, and shell better than any natural stone; we have never seen these, but have no doubt that the emery will keep sharp and stand the friction of shelling well enough; yet it might be more difficult to re-face such stones equally and truly with the emery, than to sharpen and true the natural stone with the pick; besides the natural stone answers well enough, and its first cost is not more than that of the artificial. We have seen such a wood and composition stone tried for making pearl barley, not because it was cheaper or better, but because of the danger of the natural stone being thrown into pieces by the centrifugal force, two stones having thus exploded in that same mill within three months. (See the article on barley mills.) No such reason exists for making the oatmeal mill-stones in this way, as they can be banded with iron around their circumference, for safety, while the principle of the barley mill does not admit of such bands.

The oatmeal sifter is perhaps three and a half feet long, and two and a half feet wide, generally with three sieves of tin or zinc punched with round holes of a suitable size, and space enough between the holes for the long hulls to slide along, without being canted up endways and passing through. These sieves are placed a few inches apart, one above the other in the same frame, each having an outlet for the bran or hulls at one end. This bran is sometimes caught in a spout and passed through a small fan to clean out and save the broken kernels which were too coarse to pass through the sieves; this is necessary only when the stones are imperfectly faced or hung. The bran sifted out has a portion of dust among it, which makes it good feed for animals, and a palatable and nutritious food is extracted from it by soaking in water, and straining it through a cloth or fine sieve; the liquid thus obtained is then boiled and has the appearance and consistency of cooked corn starch, but is of a bluish color and pleasant sourish taste, and is called *sowens*. The dust sifted out of the shellings is poor feed, and if it were not for the points of grain that are broken off in the process, it would be nearly worthless. The shelling is worthless, except for fuel, or for packing and protecting eggs, etc.

A valuable improvement has been made in the construction of the kiln for drying oats, by substituting for the old kiln a set of sheet-iron cylinders on the principle of the bolting reel. The iron covering is punched full of holes to facilitate the ventilation and escape of vapor, and the reels are inclosed in a brick furnace or oven, under which the fire is made, and through which it circulates. The cylinders are placed one above another, and revolve by gearing on the ends of their central shafts, which project through the brickwork for that purpose. The reels are hung with an incline in such a position that the grain fed into the high head of the upper one is passed gradually along the interior and drops into the high end of the next one, through which it passes back in the opposite direction, and is either let out dried, if the two cylinders are long enough, and the motion slow enough, or drops into another cylinder and is again traversed through the length of the kiln. Thus it will be seen that the number of cylinders and their length must be proportioned to the velocity at which the grain passes through, and the heat used, so that the oats will be just dried aright when they drop from the last cylinder. Simple arrangements are made to vary the declivity of the cylinders and the rate at which they revolve, and also to regulate and control the degree of heat.

The advantages of this plan over the old kiln are, that every part of every grain is dried equally, and in the shortest possible time, while in the old kiln some of the oats were constantly in contact with the hot plates and got scorched, while the greater part never touched the kiln at all. A further advantage is that the damp oats enter the upper cylinder, and the moisture is gradually expelled as they progress lower and nearer the fire, until they approach the lower end of the last cylinder immediately over the fire, where they fall out dry. To understand the full benefit of the gradual increase of heat thus applied, place a handful of damp oats on a hot shovel, or griddle, and many of the kernels will pop open being burst by the sudden expansion of the moisture within; but if the heat be raised and applied gradually, the same oats will stand any degree of heat, up to the point of scorching, without injury.

When steam is used as the motive power for working an oatmeal mill, the heat and smoke, after leaving the boiler, together with the exhausted steam, can be applied to heat the drying kiln. A brother of ours who owns the Barre Mills, near Lacrosse, Wisconsin, sent us a description with a plan and section of a kiln to be built over the boilers and furnaces, supplying the heat and steam in the manner indicated. The furnaces to discharge the remaining heat and smoke into the inclosed space under the kiln head, which in this case is not perforated, but steam-tight, the smoke, &c., circulates under the iron of the kiln head, which it heats and is then collected in a flue along the back side of the kiln close to the iron, and conducted into the chimney. This plan combines economy in the expense of construction, with an equal economy in the expense of working, and will, no doubt, come into general use where steam is the motive power.

Manufacture of Split Peas.

The manufacture of split peas of commerce is another branch of the milling business closely allied to the making of oatmeal and pearl barley.

The mysteries of this process are still less known and practised in the United States than either of the others, and we can think of some parties who would like the mystery to be continued, and will not like to see it dissipated by our publication of the process and machinery by which the peas are split and hulled. But in treating of all the various subjects, so far, we have withheld nothing essential that occurred to us at the time of writing, and know of no good reason why we should suppress information on this particular branch of manufacture.

The first part of the process consists of soaking the peas in a tank of cold water, or water slightly tepid, if the weather be cold. This must be continued until the farinaceous part within the hull is moistened and swelled, when the hulls being oily and less affected by the absorption of moisture, will burst and be loosened by the unequal expansion. The water is then drained off, and the peas elevated to a floor where they are spread out until the superfluous water is dried off, when they are afterwards thoroughly dried in a kiln.

This drying must be accomplished without contact with smoke, or the color and flavor of the grist will be injured. When split peas are made in connection with oatmeal, the drying is generally effected by hurrying a batch of oats from the hot kiln and withdrawing the remaining fire; the peas are then spread upon the kiln, and turned and shifted round until sufficiently dried by the remaining heat in the kiln. Sometimes cylinder driers are used for this purpose; these are a kind of *cross* between the cylinder oat kiln drier before described, and that used for roasting coffee. After being dried and cooled, the peas are split and hulled in the shelling stones which finishes the process, except that the hulls must be blown out.

When split peas are made apart from the oatmeal business, they are sometimes split and hulled between a conical cylinder and case, made of strong sheet iron and punched, the rough faces placed together and the peas passing down between these, the space being enlarged or contracted by raising or lowering the revolving cone. Another plan we have seen used for splitting peas, and hulling buckwheat, is a stone, like a barley mill-stone, or thick grinding stone, and hung like these on a horizontal shaft. It has no case round it, but only a concave made of similar stone, and resembling the water trough under a grindstone; this incloses one-fourth or more of the circumference of the stone, and is hung in an adjustable frame, one end having a permanent axis, and the other being set by a screw, either closer or further from the stone, as required. The motion of the stone draws the peas in at the movable end of the trough, and throws them out split at the other end on to a small sieve, which lets through any small fragments and saves them. A small fan then blows out the hulls, and the peas are ready for market.

This stone and its concave are both picked in small lines, commencing at each edge and running obliquely to the centre, where they meet; those cut in the stone with the wide end of their triangle foremost, and those in the concave, in the opposite direction. This arrangement of the lines gathers the peas toward the centre where they are thrown out in a round stream.

CHAPTER XVIII.

THE BARLEY MILL.

THIS is a machine for removing the hulls and tough bran or skin in which the grains of barley are enveloped, leaving the kernel clean and white, in which condition it resembles the hulled rice of commerce, and is called pearl or pot barley.

The barley mill or machine by which this is accomplished consists of a simple stone hung on its edge upon a horizontal spindle, like a grindstone. The stone used is a sharp, soft, and rough sandstone, similar to those used for hulling the oats in the manufacture of oatmeal. It is made from eight to sixteen inches thick, and from three to four feet in diameter, according to the fancy of the millwright or proprietor. The stone is inclosed in a case of sheet iron perforated with holes to let out the dust as it wears off, and not large enough to let the grain through. In punching the iron a rough burr is formed around the under edge of the holes; this rough side is placed inwards, and assists the stones very materially in the process of scouring the grain.

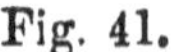

Fig. 41.

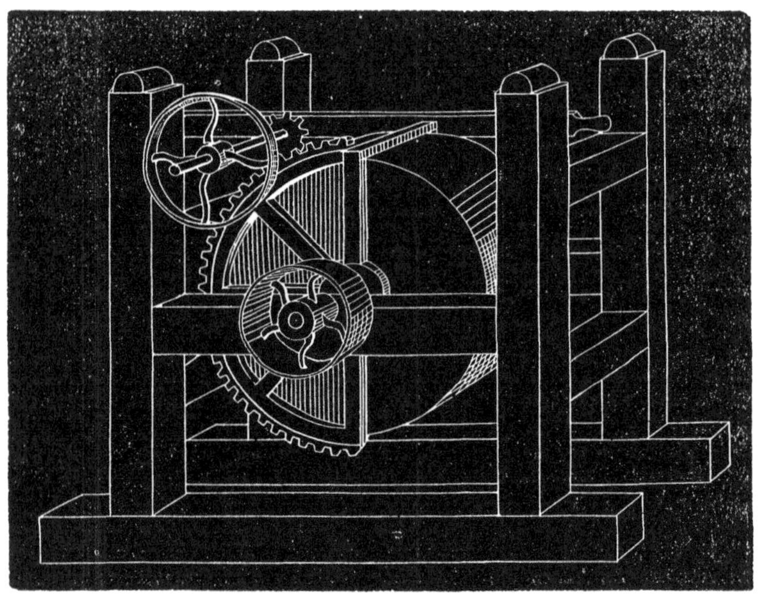

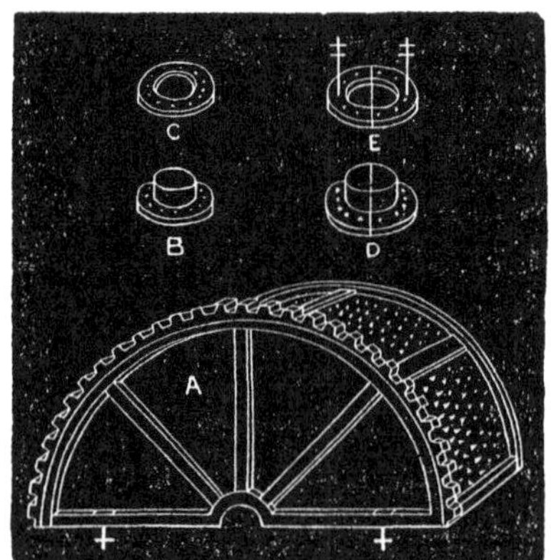

A, one-half of case, which is in two parts; *B*, flange for driving end; *C*, collar for driving end; *D*, flange for feed end; *E*, collar for feed end.

The case is placed about three-fourths of an inch clear of the stone at each side, and an inch clear round the circumference, and the contents of this space is the quantity put in for each charge or grist. The stone is geared to revolve as fast as it is considered safe to run it without throwing it to pieces by the centrifugal force; this velocity must be graduated with regard to the diameter and texture of the stone, an average being about five hundred revolutions per minute. A strong compact stone of small diameter will bear to be geared to a higher working velocity than this, but a soft tender stone, especially if it be of large diameter, should not be geared to work as high as five hundred, because allowance must be made for the increase of speed when the stone runs empty, which it is sure to do some time or other in spite of every precaution, and the utmost vigilance.

The case revolves in the *same direction* as the stone, but very slowly; it has a spur gear of segments around the circumference at one corner, and a small pinion working into this gear is connected with the driving shaft by a belt or otherwise, so as to drive, or rather *hold back*, to a velocity of between one and two feet per second.

The case is made in halves to admit of taking off and replacing without disturbing the stone or spindle. A semicircle, of a radius equal to that of the stone, and one inch added for clearance space, is made of good well seasoned hardwood, not liable to warp, and two inches thick. Each semicircular frame is of a half-circle three inches wide, with a base piece the same width running across the centre, and joining its two ends, and by bolting these base pieces of two semicircles together the whole circle is completed, or the two halves of the case fastened together. These base or centre pieces must be made wide at the centre in the form of a large circular hub to admit of a hole being cut out of each side for the spindle to pass through, and around which the flanges or ledges of iron are fastened, upon which the case hangs and turns. Two or three pieces, according to the length or number of felloes composing the outside circle, three inches wide, must be mortised in between the hub and outer circle, like spokes or arms in a wheel. These are necessary for strength and also to splice and nail the sheet iron lining to. The hole through the ends of the case referred to, should be about five inches at the end next the driving pulley, and eight or nine in diameter at the other end, as a scoop-shaped spout is fixed in this opening to feed the grain.

The ledge or flange upon which the case hangs and turns, is of the same size as these holes, around which they are placed; they are of cast iron, in two halves to admit of each being fastened to its respective half of the case; the portion projecting horizontally, which makes the bearing, is about two inches wide, and should be turned around the outside and edge; the other portion turns out at a right angle with this bearing, and so fits flat upon the outer surface of the ends of the case. It is also about two inches wide, with screw-nail holes through it to fasten it on; it should be sunk into the wood its own thickness, to help the screws and insure its stability.

The boxes or bearings in which these turn, are made of hard wood, about three inches thick, the recess for the bearing cut out of the solid wood to the proper depth and circle to fit the flanges, that is, two inches deep, and to the circle of the outer turned circumference or bearings of the flanges. These bearing boxes only require to be a semicircle below, the weight of the case being always sufficient to keep it steady and secure in its place. The boxes are sometimes bolted on to the inside of the beam or timber of the frame upon which the spindle rests, but it is better to have that beam of the frame close enough to admit of dovetailing the boxes into it three-fourths or an inch deep, and fixing a key below, and one at each end of the box, by which means it can be shifted very easily and accurately, and also held safely and securely as it is placed. This easy control of the boxes is found to be a great convenience in adjusting the case to be perfectly true with the stone; and upon this the perfection of the work greatly depends.

Another item requiring strict and careful attention, is the manner of closing the space between the spindle of the stone, and the flange bearing upon which the case turns. This is sometimes closed by metallic collars, but the spindle turns at a great velocity, and being of considerable size, never less than three inches in diameter, the friction is so great that these collars wear and cut somewhere or other, in spite of every precaution; the spindle, being horizontal, does not admit of a supply of oil being kept around it, and the collars are so much exposed to the barley dust, that oil applied to them is soon thickened and dried up.

Experience has taught us that the following arrangement is best: Suppose the opening in the end next the driving pulley to be five inches, then have a collar made by bending a piece of iron one and a quarter inch wide, and one quarter thick, edgeways, seven and a half outside diameter. This will allow it to reach out between the bearing flange and the wood three-fourths of an inch all around, and the wood must be cut away under the iron to admit it into this situation. This collar must have small holes drilled through it all around its inner edge, and about half an inch apart, and the sides and outer edge should be turned smooth and true to allow it to revolve freely in its recess without admitting the grain between.

Now take a piece of thick, strong stocking leg, about three inches long, or a piece of new knitted stuff of the same shape, made for the purpose, and sew one end of it around the inner edge of the collar with a thick woollen cord made by doubling and twisting several strands of yarn together; this is done by drawing the woollen cord through the knitted stuff and the holes drilled through the collar by a large needle. This should be sewed on to the collar so that the loose end will stand in towards the stone, and when slipped on to the spindle and in its place, if it is not sufficiently tight or elastic to prevent the grain from working out, it must be tightened by passing a piece of the woollen cord several times round it, drawing it through the knitted stuff at intervals to prevent it slipping off. The space referred to, when fitted up in this way with a good piece of knitted woollen fabric, will last for several years, requiring no care or attention, as the friction and dust appear to fill it up and thicken it rather than to wear it, and it never wears or cuts the spindle like the iron or brass collars.

Those millwrights who use metallic collars to close the space referred to in the back end of the case, generally close that in the front end in the same way, and make a small hole through, which is furnished with a small spout spread out in the form of a hopper at the top to feed in the grain. These collars are liable to all the objections urged against those at the other end, besides another very serious one, which is, that they close up the centre and exclude the cool air, a strong draught of which is essential to keep the interior cool, and should be admitted here. The open space in the front end should not be less than eight or nine inches in diameter; this should be furnished with an open ring or collar like the other, its outside diameter an inch and a half more than the inside of the bearing flange supporting the case, behind which it is placed in a recess cut out of the wood to admit it, like the other. It should also be turned off true around the sides and outer edge, and drilled around the inner edge for rivets, about three-fourths of an inch apart; this one should be of heavier iron than that we proposed for the other end, because it must have two standards welded solidly into it at opposite sides, and standing out at right angles from its face, and long enough when it is in its place to reach out and pass through a projecting lug standing up from each end of the cast box upon which the spindle turns. These lugs upon the box and the diameter of the collar must correspond, so that the standards may pass straight and level through the holes in the lugs; the standards are screwed at the end, and two nuts placed upon each, one behind and the other before the lug referred to, and by turning these nuts one way or the other, the collar is adjusted to its place and fastened. The standards should also be made to bear near the collar upon the edge of the wooden box which supports the flange and case; this support being near the root of the standard, secures the necessary stability to the collar.

This collar, instead of having the piece of woollen stuff to close the space, has a zinc spout riveted on with copper rivets around its inside; this spout should rise up at an angle of about forty-five degrees. It should be a whole spout down next to the case, that is, it should be riveted on all around the inside of the collar, with the seam up; but the top may be cut out, commencing about four inches up from the case and scolloping it with an easy curve upward until only the under semicircle is left, and this may be rounded off gradually to the end like a meal scoop, at the height of about two feet from the collar. This can generally be still more shortened by observing how far the barley works up along the spout when the machine is operating, when it may be cut off at a few inches beyond the point at which the grain arrives, which will shorten the scoop to about fifteen inches.

It is important to have the scoop as short as possible, as the miller needs to put his hand down into it and take out some of the barley to see how it is progressing, and when it is thoroughly made and fit to be taken out. The machine is charged through this scoop, the best way being to elevate the grain into a store hopper over head, and let it down through a spout by opening a small valve or gate; but it can also be filled into this scoop by hand. This scoop spout remaining always open, admits a current of cool air, which is drawn into the interior by the centrifugal force engendered by the rapid motion of the stone, and counteracts the tendency to heat, which the spindle, stone, grain, and everything inside of the case has in a degree, from the great amount of friction.

A small door is made through the outside circumference of the case to let out the charge of barley when it is finished; this door is hung by hinges at one edge, and when closed the other edge is secured by two buttons. To unload a charge, a box or drawer is slipped in under the machine on the floor, the small pinion controlling the case is thrown out of gear by a lever fixed for that purpose, and the case is turned by hand until the door is below, when it is opened and the case is moved backward and forward, until it is emptied.

The little door, as well as the whole circumference of the case, requires to be strong, as the grain occasionally is liable to pack between it and the stone, so suddenly and solidly that it will stop the whole machinery, water-wheel and all, as short as if it ran against a stump. For this reason the iron covering of the case requires to be secured and strengthened by strips of tough wood, two inches wide and one thick, bent round each corner, and strongly fastened with screws. Cross-bars are also put across between these bent pieces, at equal distances, their ends being held by these, and the iron is nailed to these bars from the inside, before the two halves of the case are put together.

The process of punching the iron by hand is slow and tedious, and it requires to be carefully done to avoid stretching and puckering the sheets out of shape. The iron should be regularly laid out and marked by a straight edge, with a red lead pencil, both lengthwise and across the sheet, and the punch driven through the angle formed by the crossing of the lines. The holes should be placed in lines zigzag, that is, the holes in one line should be intermediate between those in the adjoining line, and thus break joints. Some millwrights make small round or three square holes, and quite close together; we prefer making the holes oblong, from three-eighths to half an inch long, and as wide as may be to prevent the grain passing through.

The punch for making these holes should be carefully made and tempered, in order to cut the piece out clean and equal, as sharp, ragged, unequal projections cut, scratch, and waste the kernels of grain, besides spoiling the appearance of the finished barley. This is the reason, we conceive, why a case punched with these long openings, their length of course running crosswise, is found to make better work than those with round or square ones. By better work we mean that the grain is cleaned with less reduction of size and weight, and retains the natural oblong form of the kernel better. The punch should be filed or ground very square to the size of the intended holes, and also back parallel each way to the same size as the point, as far as it will be driven through the iron; this is necessary to avoid bulging and stretching the iron out of shape. Some punch the sheet clear to the edge to avoid puckering it, but it is stronger and better to leave a selvage as wide as the wood upon which it is to be nailed; if carefully punched as here directed, a little hammering of the selvage upon an anvil will bring all right.

The punching may be done upon smooth blocks of hard wood, endways, or upon a block of lead, and it is good economy for the workman to practise at the commencement until he can drive the punch clean through at one stroke of the hammer. He will do it so much better, and more expeditiously, that he will soon make up the time spent in acquiring the proper sleight and weight of stroke. When the surface of the wood becomes too much damaged, the end must be sawed off and dressed anew with a plane; when that occurs with the lead, it may be hammered smooth again.

In some of the last and best working mills that we have fitted up, the iron lining the ends of the case was not punched at all. The reason of this will be explained when we come to consider the working of the machine. We will try to give a description of a barley mill which we fitted up for John McKenzie, of Burke, N. Y., it being the last one we wrought upon, except some alterations made upon one in Ogdensburg, N. Y., last summer.

McKenzie's new mill is a double one, and consists of two distinct and separate machines, placed side by side, and parallel with each other, the driving pulleys of both being inward, and meeting so closely and exactly that they resemble a true and false pulley upon the same shaft. The driving belt comes up around these from the story below, from a small drum, which is also double, having two swelled tracks for a belt, one at each end, corresponding with the two pulleys on the machine above. This belt is laced pretty slack, and tightened to the working grip by a tightener that lies upon it, the tightener also being of double length, but parallel throughout, as the tightener should always be.

When one machine is working and the charge it contains is finished, the miller withdraws the tightener to slack the belt, and shifts it over on to the pulley of the other machine. The tightener is then let on again, and the other machine proceeds to work off its charge, while the miller is emptying out the finished charge of the first machine, and refilling it to be ready when that in the other machine is completed. By thus shifting from one to the other alternately, one is kept constantly working while the other is at rest, and the finished charge is being withdrawn and replaced by a new batch to be ready. This arrangement has another advantage over the single machine, which is, that a little time is gained at each change, for the machine to stand and cool; the making of a batch requiring about twenty minutes, while unloading and refilling only occupies about five.

These two machines are exactly alike, except that they both face outward, the driving pulleys being together.

The stones are thirty-seven inches in diameter, and fifteen inches thick. The cases were placed three-fourths of an inch clear of the stone, at each end, and an inch clear around the circumference; but the space gradually enlarges as the stone wears, and is dressed off with the mill pick to keep it true. The bridge beams upon which the spindle rests in the frame, are twenty-four and a half inches apart; that is, fifteen inches for the thickness of the stone, one and a half for clearance, four inches for the thickness of the two sides of the case, and two inches clearance for the box bearings and flanges which support the case each side. These boxes are three inches thick, and are dovetailed into the bridge beams one inch deep. The upper surfaces of the bridge beams, where the spindle bearings are placed, are thirty-two inches from the floor, which leaves about a foot of clearance between the case and floor for the box or drawer to be slipped under, to receive the charge of barley.

The length of the frame is just sufficient to clear the cases comfortably; the corner posts reach up a little higher than the top of the case, and a small beam is mortised in between each pair of *inside* posts near the top, above and parallel with the bridge beams; in these are placed the bearings for the shaft of the small pinions which mesh into the gearing around the cases. A large pulley is placed upon this shaft, between these two bearings, and a belt connects it with a small sheave on the drum-shaft below, to give it the requisite slow motion. The gearing is on the *outside* of both cases, and a universal joint at the bearing of the shaft referred to, joins another piece to it at each end, upon which the pinions are placed. These jointed pieces of shaft have each a bearing in a lever outside of their respective pinions, by which they are thrown in and out of gear. The pinion is placed near the top of the case, a little behind the centre, and the back end of the controlling lever is mortised through the top of the back post of the frame, and a pin put through the end. This pin is the fulcrum or hinge upon which that end of the lever works. The other end is rounded down to a convenient size and shape to take hold of to move it. When this small end of the lever is brought down under a pin or other catch, the pinion is held *in* gear; when thrown up above that catch it is held *out* of gear.

The frames of these machines are made of spruce, eight inches square; the tenons are supplemented with heavy joint bolts, as it is essential that the frames and floor upon which they stand should be perfectly solid. The iron forming the circumference of the cases is punched with holes three-eighths of an inch long, the rows being that same distance apart both ways. For convenience in laying these out, we stepped the distance on a strip of wood the length of the sheets and marked both edges and both ends by this, and ruled across from mark to mark both ways, using a soft red lead pencil.

The iron lining the ends is not punched at all. The iron is nailed on to the wooden frames with small wrought clout nails (tacks) five-eighths or three-fourths of an inch long; they should not be longer than three-quarters, as the iron will wear out before the wooden framework of the cases, and the nails should be of a length that would admit of their being drawn when the iron has to be renewed.

The spindles upon which the stones are hung are round, three inches and a half in diameter, with a cast-iron block five inches square and fourteen inches long keyed on, upon which the stone is placed. The keys should run clear through the blocks, and the key-way in the shaft should be cut a little longer than the blocks, to admit of shifting them either way to get everything to agree. The eyes through the stones are five and a half inches square. To get the stone true and secure upon the block, we placed the spindle upon two bearings upon which it could turn freely, with the stone upon it; we then made eight long thin wedges of dry wood, an inch and a half wide, and entered two of these in each square between the cast-iron box and the stone, one with the head each way, all around; then turned the stone round, marking the high places with a piece of chalk held firmly upon a rest, and drove the wedges in, or slacked others back, as the case required, until the stone was true. We closed the opening all around at one end, and turned the other end up, and filled the rest of the space between the stone and box full of melted lead.

It is not possible for a miller to dress such stones perfectly true before they are hung, and for that reason he should take a bar of iron, and by holding the end against the stone and over a solid rest, turn a number of rings or creases a little distance apart all around the stone, after it is set in motion. With these turned creases for a guide, he can dress off the intermediate spaces sufficiently true with the mill pick. Some millers dress the stone with a pointed pick, such as is used upon an oat shelling stone; others crack them with a flat pick like a flouring stone, making the cracks to run across the stone around the circumference, and from that to the eye on both sides. This is done under the impression that it will hull the barley quicker and better; but, as far as our experience goes, such is not the case, and to rough the stone in this way has an effect upon the grain similar to that of the sharp points inside of the case made by punching the iron with a round or square point already referred to, that is, the grain is wasted and worn off, more particularly the *ends* of the kernels, which if made in a machine with both case and stone made rough in this way, will be worn off so as to look like small peas before the hull is all cleaned out of the little crease (eye) running down the side of the kernel.

This not only makes a great waste, but it spoils the appearance of the barley, and is caused by the little corners or angles made by the pick catching the ends of the kernels next to the stone and canting them over endways like the oats in the process of shelling. To avoid this, the stone should be dressed as true as possible, and depend upon the sharp sandy natural grit of the stone for scouring; the pick should never afterwards be used upon the stone except to reduce any high hard spots, and break the glaze upon the surface when it occurs.

In addition to this, it is necessary in order to insure the best of work, that the space between the stone and case should be kept packed as full as possible all the time the machine is working; and in order to keep it so, it must be filled up once at least (twice is better) during the making of a batch. For this purpose a supply of barley partly made, should be kept convenient to fill in, when it becomes slack by the dust sifting through the case. It is well known to barley millers that the tighter it is kept packed in this way, the less waste will occur in making, and the better the barley will retain its natural shape. The hard outside hulls of the grain perform an important part in the process of scouring; these, when examined by a microscope, are seen to be serrated along the outside edges, like a new file or very fine saw-teeth, and being quite hard and sharp, they assist very materially in rasping off the thin tough cuticle which envelops the kernel inside of the rough hulls. If these little rasping hulls were blown out, or otherwise separated from the mass as soon as they are loosened from the grain, the process of scouring would neither be so well nor so expeditiously done; as these answer the same purpose here that the emery does in the revolving iron cylinders in which the brass clock machinery, iron gimlet handles, and a hundred other similar small articles, are smoothed and polished.

We have alluded to the injurious effect of allowing the grain to get slack as the making progresses, that is, an empty space forming at the top of the stone by the diminution of the bulk of the charge, as the dust is worn off and sifted out through the case. At this empty space, the grain being unconfined is made to revolve or spin round rapidly by the friction and velocity of the stone, and the ends being weaker and more exposed than the middle of the kernel, are rounded off, and it is reduced to the form and appearance of a small white pea, before the seam in the side is thoroughly cleaned out. The motion of the case and stone keeps the grain continually shifting, so that the whole charge is soon subjected to this triturating process unless the empty space be filled up. The effect upon the barley is very similar to that referred to as being produced by the sharp angles sometimes made in the surface of the stone by the pick.

To obviate this difficulty, the most improved machines are made with the stones of small diameter and greater thickness, and consequently breadth of circumference, to make up the necessary working surface. This admits of running the machine much more densely packed with grain than when the stone is of large diameter and thin, besides diminishing the amount of end face, and increasing that of the circumference is a profitable exchange, as the principal part of the work is performed by the circumference, and but little by the ends, particularly near the centre.

Of several barley mills that have been fitted up within the last three or four years, the smallest diameter of stone used that we have known of, is three feet, and fifteen and a half inches thick; the largest is five feet in diameter, and ten inches thick. Several are from four to four and a half feet in diameter, and only eight or nine inches thick. We have examined the working of all these at various times, and the result is invariably as indicated above, that is, the smallest diameter and broadest circumference face does the best work, the barley being finished more uniformly, with less waste.

Uniformity in size and appearance of the kernels of pearl barley is desirable, but cannot be attained with any barley as it comes from the threshing machine, for the reason that the grains do not grow equal in size, neither are they equal in shape or hardness, and the mill cannot equalize these differences. The different grades must therefore be separated by screening. This is done by passing it slowly through a screen composed of four different sheets of zinc, each punched with a different size of holes—the coarsest being uppermost takes out any foreign matter larger than the barley, and a few overgrown kernels of it—the two middle screens make the different grades of barley, and the lower one lets through all grains that are too small, and also broken pieces. This screen is best driven by a double crank, one on each end of the shaft, and of very short and rapid stroke. The grain should be passed through a screen as it is delivered from the elevators into the stone hopper above the machine; and also subjected to the action of a suction or other fan to clean out the dust, &c., before it is let into the machine.

On the question whether to punch the iron, lining of the ends of the case, or leave it plain, our experience has not been sufficient to enable us to decide, or scarcely to offer an opinion. It is claimed by the advocates of the smooth iron that the barley is made with less waste, and retains its natural form better; on the other side, it must be admitted that it will make faster and keep cooler by having the iron punched.

We have alluded to the danger of the stone exploding by the accumulation of centrifugal force, when the velocity of the stone is very much increased, by its running empty or by other accident. Several instances of this have occurred lately. In one case a boy was left to attend a run of flour stones and barley mill at night—the same undershot wheel drove both—it was geared by a pit-wheel on the water-wheel shaft, working in a crown-wheel upon a spur-wheel shaft. The spur-wheel was ten feet diameter, and drove four run of burr stones, besides the smut machine and this barley mill. The boy fell asleep, and the flour stone ran empty; the charge of barley soon became so much reduced by wearing and sifting through the case that it offered but little resistance to the power upon the wheel, and there was so much gearing intervening between the wheel and the barley mill, each set gaining speed, that the velocity of the barley-mill soon increased until the centrifugal force overcame the cohesion of the stone, and it parted in three pieces.

The stone was about fifty inches in diameter, and was drilled through from the circumference to the eye at three equi-distant places, and three iron bolts put through these holes to strengthen it, the flat heads of the bolts being countersunk into the stone at the outside, and the points tightened in the iron box at the centre by a screw nut. The stone parted at these three bolts—one piece passed obliquely up through the floor into the story above, cutting its way completely through a large beam over the boy's head, and waking him up; another piece passed down through the floor into the basement, and the third piece went out through the stone wall of the mill into the river.

The next accident of this kind we have to record occurred in McKenzie's celebrated barley mill, at Burke, near Malone, Franklin Co., N. Y. It is the same double mill described in detail in this chapter, and was built as an addition to, and adjoining a grist-mill, of three run of burr stones, which does an extensive business both in custom and manufacturing work. It is upon a small but durable stream, and in order to economize the water and other expenses both were driven by the same overshot water wheel, the barley mill being worked only at night after the day's work in the grist-mill was finished. In consequence of this arrangement the barley miller could only sleep in the day-time, and frequently did not get sufficient sleep; at all events, he fell asleep while one machine was running. This mill, like that driven by the undershot, had so much intervening machinery between it and the water wheel that it multiplied the velocity to such a degree as the case worked empty, that when the miller awoke the stone was humming like a circular saw. McKenzie, who is posted in these matters, and the best barley miller that we know of, had frequently warned him, that in case of falling asleep, or otherwise letting the stone run away, never to shut the water-gate down suddenly, but throw in a little grain and check the water a little, alternately, and gradually, until the speed was reduced to a safe rate, and then stop the machinery altogether.

But the miller being stupefied by sleep, and frightened, did not think of the warning, but ran to the gate, which was at the further side of the grist-mill, and slammed it down. This checked the water wheel suddenly, which, being the slowest mover and the driving power, had gradually accelerated the velocity of the intervening machinery and stone, until the latter had attained the maximum velocity, and was running in unison with the rest of the machinery. The wheel checked the intermediate machinery up to the stone, which, being the swiftest mover, and very heavy, had accumulated a tremendous velocity and momentum as a fly wheel, and now suddenly became the driving power. The shock occasioned by the back-lash of this reversal overcame the little remaining cohesion of the stone, and it parted into four equal pieces, from the four corners of the eye. It demolished the case and frame, of course, but damaged the building less than might have been expected; one quarter passed out through the end of the mill, and over the river, landing on the opposite bank.

The miller was so frightened and chagrined that he would not attend the other machine any longer, but exchanged places with another young man who attended the grist-mill, and who was conceited enough to think that the like never would have taken place with him. He was doomed, however, to have this conceit very suddenly taken out of him; for when he had run the single machine about six weeks or two months he fell asleep at his post, and the same thing occurred with him—every circumstance being almost identical with the previous accident. It is astonishing and difficult to account for the infatuation that could impel him, when he awoke and found that the mill had run away, to run to the gate and shut it down the first thing, especially after he had heard the philosophy of the whole occurrence discussed and explained, and been warned against the *immediate cause* of the other accident, to wit, the back-lash by the sudden check of the machinery. But, by his own account, he ran directly to the gate and shut it down; and, before he had time to move from where he stood, or take his hand off the gate lever, the explosion occurred.

The second accident caused considerably more damage to the building than the first one; one quarter of the stone struck the main beam in the end of the building and smashed it out; another broke through the lower floor, and the case and frame were broken in fragments. The interior of the building presented the appearance of having been struck by lightning, or blown up with gunpowder, and had the accident occurred in daylight, with the mill full of people, as it frequently was, the damage to life and limb might have been very serious.

When this second and last stone went to pieces, McKenzie tried the experiment of a composition stone. For this purpose, he had one made of dry hard wood; it was made of disks or layers of two inch plank, well jointed, and pinned through, the joints being filled and the pins driven with white lead for cement. The outside was then turned off to the size and shape of the other stones and covered over with several coats of glue and emery. This was put on in a severe frosty time, and in a hurry, the "stone" being placed near the mill stove to keep it from freezing, and hasten the drying; but parts of the composition were frozen before drying, and other parts overheated and dried too suddenly.

It was hung in place and set to work immediately, and found to hull and scour the barley faster and better than any natural stone, but when examined after running a while, portions of the composition were scaled off. It was afterwards coated anew, the emery this time fastened by a "patent" waterproof glue; this was applied to the stone in its place late in the afternoon, and the next morning it was set to work. Although hot irons were placed under and around the stone to hasten the drying, still it was not so dry as it should have been when set to work, and the consequence was that a good deal of the composition again wore off.

The manufacture being depressed, no further attempt was made to run it, and McKenzie has sent to Scotland for two more stones from the same quarry; and this summer (1869), having a new building and water-power fitted up a little below the grist-mill, which will use the same water over again, he intends to transfer the barley mills into this and run them with a small turbine wheel, geared in such a way that the velocity of the stone will be but little accelerated when running empty.

We were sorry that McKenzie should have abandoned this experiment without giving it a fairer chance of trial, as from what we saw of the working of the artificial stone, we were satisfied that it would work faster and better than any natural stone that can be found, and still believe that if the composition were applied in warm weather, and sufficient time was allowed for it to dry gradually and thoroughly before it was set to work, it would stand the friction well. We have often made and used emery wheels and cylinders for many different purposes, where the grinding and abrasion were all confined to one spot, or to one side of the cylinder, as in grinding cast-iron rollers, or machine cards, and found it to stand the friction well; and can see no good reason why it should not stand for a barley mill-stone, where the friction was equally distributed over its whole surface, and could not be sufficient at any point to make much impression upon an emery coating carefully put on and sufficiently dried before using.

Some readers may think that we are occupying too much time and space with these details, and consider them frivolous; but for those owning, making, or running such machines they may be sufficiently interesting to require no apology, and under cover of this remark we will mention only one more accident that occurred with a barley mill, which illustrates the necessity of having the barley stone, or any other heavy machine revolving at a swift velocity, perfectly balanced.

The stone referred to was of unequal texture, one side being closer in the grain, and consequently heavier than the rest of the stone; this denser side was also harder, and wore away more slowly than the other. It was balanced by cutting a hole in the light side and running in lead, and a similar hole in the heavy side, which was filled with plaster. Although accurately balanced at first, the balance gradually varied by the soft side of the stone wearing away until it was perceptibly out of balance when running swiftly. In this condition it was left to stand idle for a while, being in gear just as it was run. During the night the rats chewed off the rope which held a waste gate open in the spout which supplied the overshot wheel; this let the whole water on to the wheel, and as it had nothing to drive but the empty barley mill, of course it drove it in a very lively manner.

It was not heard nor noticed until a girl, early in the morning, went down the ravine some distance below the mill for spring water. Here she heard a tremendous crash, and at the same time something like a streak of lightning enveloped in a small cloud, and smelling strongly of sulphur or gunpowder, which passed her so close that it threw her down, although it did not touch her. She sprang up, and looking down the ravine she could distinctly see the barley mill-stone as it was running directly away from her. She described it as running in a direct line, neither veering to the right nor to the left, but every little while jumping straight up into the air, sometimes as high as ten or twelve feet. Some of these jumps, which appeared to her to be perpendicular, were afterwards measured and found to clear from twenty to forty-three feet. After running a quarter of a mile down the hollow it came to an up grade which checked its speed, and it finally brought up against a steep bank. It was not broken, but the corners were chipped and rounded off sufficiently to preclude its ever again being used as a barley mill-stone, a purpose for which it never was fit, and never should have been employed.

The explanation of its thus deserting its post was, that its being empty, somewhat heavy on one side, and the journals and bearings all dry, the great velocity and weight soon heated the spindle and bearings, and they began to wear and cut; this gave some play to the spindle and stone, which increased the power of the heavy side for mischief, and encouraged the wear of the bearings still more. Thus the two defects, mutually helping, soon increased the swing and jolt, until the strain broke the spindle, and the stone being liberated took its departure; the accumulated rotary power within itself furnishing ample motive power for the purpose, and the friction of its weight upon the floor gave it the requisite forward motion to enable it to jump through the window and continue its course down the ravine, like a cart wheel along the road, until its rotary force and acquired headlong momentum were both gradually overcome.

It is rather singular that all these accidents should have occurred and no persons be killed or hurt by them; however, this result is due, not to the harmless nature of the accidents, which are really as much to be dreaded as the bursting of a cannon or the explosion of a bomb-shell, but to the fact of their occurring when no one happened to be in the way.

But this circumstance should not deter any one who may be interested in fitting up such machines, from using every precaution against the possibility of such an occurrence taking place. The most effectual safe-guard is to gear the stone to the wheel in such a way that in case of the stone running empty its velocity cannot be increased much beyond double its working speed. Gearing thus with some kind of wheels would involve a waste of water; where this cannot be afforded, we would advise, either using a different wheel, or other-wise providing the wheel with a reliable governor to control the speed by the water-gate.

Whether any reliable addition can be made to the strength of the stone by drilling and placing screw bolts through it, from the iron in the eye to the circum-ference, as described in the first accident, we very much doubt. It would be only reasonable to think that the hole drilled out of the solid stone would weaken it as much at that point as the iron would strengthen it; besides, the bolt cannot prevent it separating *along its course*, which the hole has rendered the weakest part, and therefore most liable to give way. And again, in case of the stone running empty, which is the only danger, the bearings become dry and the spindle and box in the eye heat and expand before the stone is heated, and iron being a better conductor than stone, the bolt is also heated and expanded in a slight degree, but is still sufficient, with the expansion of the central irons, to slacken the grip of the bolts, and leave the weakened stone to its own resources. For these reasons we have never put through any such bolts, and cannot recommend them.

There are some secrets known only to those thoroughly initiated, that are insisted on as being of importance in order to manufacture a first-class article of pearl barley, and are guarded with such care that some crack millers exact an oath of secrecy from their apprentices, in order to preserve them in their craft. All we can say is, that there is nothing in them requisite to make a really good article, but the pretended sleight consists in making perhaps a nicer looking article, although not in reality any better for use or consumption.

CHAPTER XIX.

WOOL CARDING AND CLOTH FULLING AND DRESSING

THIS business is frequently carried on in connection with gristing and other branches of the milling business, and the same millwright is required to set these in operation. The carding machines are furnished complete from factories, and the millwright has only to adapt the proper power and speed to the driving belt to set them to work, and occasionally to assist the carder to turn the cylinders anew, and fit on and grind the new cards. But the fulling mill has to be made on the spot, and be adapted to the situation, circumstances, and power by which it is to be driven, and we have seen many a good millwright sorely perplexed and puzzled to reconcile all the contingencies, and give the trough and hammers the right curves and angles required, to get the

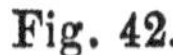

Fig. 42.

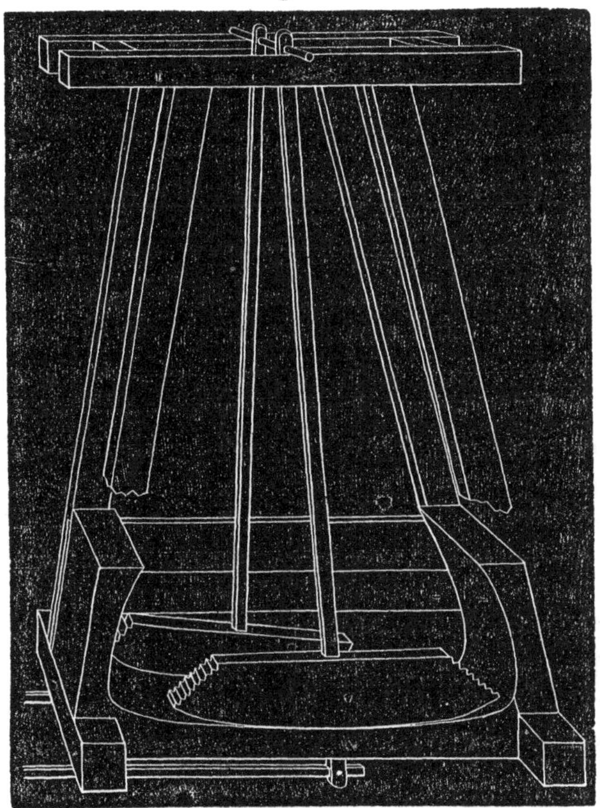

stock of cloth to turn just fast enough, and not too fast, to full equally and quickly. When the mill is made single, that is, to full only one stock at a time, this is not so difficult, because the distance of the hammers from the curved head block can be varied, and the angle at which they strike the cloth altered by shifting the axis above, upon which the handles are hung. But the double mill, for two stocks, admits of no such adjustment, as any alteration made to benefit one end alters the other end in the opposite direction, so that the benefit to one is made at the expense of the other, and the only remedy is to take it down and make the necessary alteration.

The original cause of the trouble is undoubtedly the apparent simplicity of the whole concern, which tempts a man to go right to work and make it, forgetting, or rather never noticing or knowing, that the fulling mill, like other machines, works upon fixed mechanical principles. The first and most important of these is, that the point of each hammer, as it moves, describes an arc of a circle, the radius of which is the direct distance from the point of the hammer to the axis on which the handle is slung, and the length of the arc described, or rather its chord, is the stroke of the crank; therefore the bottom of the trough must be marked and curved by this radius from the point of suspension indicated, and for a distance at least equal to the length of stroke. Behind this curve the bottom should be continued straight, and a little descending to the end if it be a single mill, or to the middle if it be a double one. The descent is required to let the scouring water run off and to ventilate.

Above this curve, at the point of the hammers, the head block should curve up rather abruptly until past the perpendicular, and then extend nearly straight, and inclining back over the points of the hammers as shown in Fig. 42.

The hammers are about 12 inches thick and 21 deep, the ends bevelled off to an angle of 40 degrees, and the incline cut in notches like saw teeth through the whole thickness. These teeth crowd the under side of the stock of cloth forward into the short curve at the foot of the headblock, up which it rises, and falls back from the overhanging upper end upon the withdrawn hammer, which shoves it up again, and the two hammers repeat the operation alternately, the cloth turning a little each time, until every part is equally and sufficiently fulled. The under sides of the hammers are circled a short distance back from the point by the same radius as the bottom, and from the same centre—the point of suspension—they are otherwise straight below.

The handles are put through a mortise in the centre of the hammers, and pinned fast; if for a double mill, the ends must pass down through the bottom of the trough, and the pitmen from the cranks be jointed on to these ends. For a single mill a slit is cut out of the back ends of the hammers to admit the ends of the pitmen, and a noddle-pin is inserted to hinge them together. The crank in either case should be placed so that the pitmen will work horizontal and parallel—the crank working freer, and keeping better, by having a bearing on both sides of the pitmen. If both are on the same side, they should be a considerable distance apart, otherwise it is difficult to keep them tight and true. We once cured a crank which was condemned, because the bearings were only eleven inches apart, and could not be kept tight, by substituting an old barley mill-stone for the driving sheave. We covered the circumference with a thick coat of plaster and glue, and turned it off. It made a good sheave, and its weight kept the short shaft steady in its bearings, besides it answered the purpose of a fly-wheel.

The first fulling mills we made we mortised the bottom piece and tenoned the head-blocks directly into it; but this cuts the wood too short across the grain at the most particular part of the curve, and it does not stand well. All we have made for many years were both bottom and head-blocks tenoned into a cross timber at each end, in which the shortest part of the curve is cut as shown. These cross sticks are made long and large enough for supporting sills, and a pair of posts are set into the projecting ends of each, these posts holding the sides of the trough in place, and also supporting the frame upon which the axis of the hammer handles is placed.

Fulling depends upon a series of microscopic barbs similar to that on a fish-hook, but continuous along the whole length of each fibre of wool or fur, and too minute to be seen by the naked eye; these admit of the fibre moving and tightening up in one direction, but, like the beards of grain, they hold all they gain in that direction. By agitating a fabric composed of such materials, as in a fulling mill, and saturated and slippery with soapsuds, the fibres are more and more entangled, and the fabric concentrated and thickened, with a corresponding contraction in length and breadth, until the texture and appearance of the cloth are changed. The principle is the same as that of felting, by which cloth is made without spinning or weaving, and the fur and woollen hat bodies are made in this way.

This peculiar tendency to condense into a thick mat is possessed more perfectly by some kinds of wool and fabrics than others, and this difference must be understood by the fuller, and modified accordingly. A piece of cloth possessing a tendency to full rapidly frequently gets into folds and wrinkles, and these, if not watched and straightened out, soon become fulled and matted as if grown together, and the cloth is spoiled. To guard against this, and full each piece aright, the fuller withdraws each stock once or oftener, stretching each piece, and leaving some out for a time to allow the others, more tardy in the process, to catch up. When the whole batch is fulled sufficiently a stream of clean water is run in among the cloth and the mill kept running until the soap is washed out and the cloth left clean, when it is taken out and hung over a pin to drain the water off, and is then stretched out on the tenter bars to dry.

Tenter Bars.

The tenter bars, on which the fulled cloth is stretched and dried, are placed either in a long open shed, through which the air has a free circulation, or as frequently in the open air. They are composed of two tiers of scantling, about four inches square, placed upon posts six feet high above ground. The upper tier of bars is fastened to the top of the posts in a straight uninterrupted line; but the other tier, the breadth of the cloth below these, is jointed together by a double tenon made on one end, and a single one on the other, which is placed between the two, with a pin inserted through these at the centre, which forms the hinge. Tenter hooks are driven into the ranges of bars, about three inches apart and the entire length, to hook the edges of the cloth on to; the hooks are of galvanized or tinned iron to prevent rust, which would stain the cloth. The lower bars are fitted to slip freely up or down on slats, four inches wide and an inch thick, which pass through mortises in each bar, opposite the posts; the upper and lower ends of these slats are fastened to the posts, the middle being left free for the bars to move upon. To place the cloth upon the tenter bars it is neatly folded (not rolled up) and carried upon the left arm, the end is hooked on to both bars at the corners, and the piece is carried along and hooked at intervals to the bars by the right hand until spread out to its full length; the operator then commences at the first end, and stretches both edges out lengthways, and fastens upon every hook; he then passes along and crowds the lower bar down, and slips a pin through it into each post, which keeps it down and stretches the width of the cloth. It is left in this position until dry—the length of time depending more on the circulation of the air, and the presence or absence of moisture in it, than on the temperature; as the cloth will dry well, although it be frozen hard all the time, provided the air be dry and circulate freely.

The Shearing Machine.

With the operation of *shearing* which the cloth undergoes the millwright has little to do, as the machine by which this is performed, like the carding machine, is furnished ready for use, and it has only to be located conveniently, and in a proper position, and the necessary motion imparted to its driving belt. The shears, however, require to be ground and sharpened occasionally, and as this is a particular job which has to be done when distant from the factory where they are made, we will try to describe the principle on which they operate, and the method which we have employed to sharpen them.

The revolving shears are plates of steel wound spirally round a small cylinder, their length being equal to the widest cloth to be shorn. These blades are sharpened on the foremost corner of the outer edge, and when working, sweep obliquely across the sharp corner of the straight blade beneath them, the progressive motion of the revolving blades along the stationary edge being similar to that of closing the blades of ordinary shears. The cloth is wound upon a roller and the end passed over a square corner directly behind the cutting edges of the shears, and fastened on to another roller on the other side; this last roller has a slow revolving motion given it which winds the cloth upon itself, and draws it from the first roller and over the corner mentioned, which keeps the cloth tight and equally up to the shears, and these clip off all the projecting wool to an equal length as it is thus presented to them.

The *ordinary* sharpening of the shears is performed by running them backwards and applying flour of emery and oil to the passing edges with a rag or brush; by this method of sharpening and the wear of working, the blades get to be uneven on the edges, and require to be ground true. To grind these shears a stone of fine and equal grit is required; it must be hung true and firm in its bearings, and the circumference turned off true; a thin stone answers best. The shears must also be hung in a temporary frame, with true guides to keep it parallel with the stone, and at the same time admit of moving endways the whole length of the blades; the stone should have a pretty swift motion, and the shears a slow motion *in the same direction*, and the shears must be kept slowly moving from end to end, to insure an equal and true edge. When this is accomplished, the stationary blade must be trued the same way, and when put in place they must be ground together with emery and oil as before to complete the fit; after grinding with emery the edges should always be finished by rubbing both angles lengthways with a torquois or Washita oil-stone. It is always best to send the shears to the factory for grinding when it can conveniently be done.

The Cloth Press.

After the cloth is sheared it is curried over with a card and brush; these are driven over it always in the same direction, in order to lay the incline of the nap or projecting wool all one way. The cloth is then carefully folded with a press-board of smooth paper pasteboard between; plates of heated cast iron are also placed at intervals between the folds, and the mass sometimes still further heated by being placed on a thick cast-iron plate, with a furnace under, in which fire is placed. In this situation it is subjected to a great pressure by a strong screw, and left under that pressure for a considerable time, the screw being tightened occasionally during the interval; the pressure and heat, with the smooth press-boards intervened between the folds, give the cloth a smooth and glossy finish, and improve the texture as well as the appearance.

The structure and fitting up of the cloth press are simple; the frame consists of two strong posts of hard wood about eight feet long, with two equally strong beams framed across between with double tenons, the top of the lower about twenty inches above the floor; the upper beam, in the centre of which the nut of the screw is sunk, is placed a little more than the length of the screw above the other one. A strong cast-iron socket or follower is fixed under the round pivot of the screw, to which it is attached by a swivel. This iron follower is bolted to a wooden one of strong plank, with a tenon on each end. These tenons traverse up and down in grooves cut in the posts for that purpose; this allows the plank to follow the screw up or down, but prevents it from turning as the screw is turned. Two holes through, near the butt end of the screw, at right angles with each other, admit of a bar being thrust through, by which it is worked.

With the coloring or dyeing, which is a chemical business, the millwright has little to do except to fix curbs and conveniences around the copper boilers, and suitable reels for overhauling and handling the hot cloth, and for these, the manufacturer himself will give better directions than we can.

The same remark might apply to our proposed directions for turning the card cylinders, and grinding and pointing the teeth of the new cards; but as the process of turning these cylinders and that of turning surfaces true by an emery wheel or cylinder, are applicable to many other purposes than those under consideration, we will try to give an idea of both here.

To turn such cylinders true from one end to the other, it is necessary to have a gauge or guide to control the turning tool, that is true, and also a tool that is adapted to the control of such a guide. The guide is made of a piece of clear and dry scantling about four inches square, and long enough to reach and rest upon the frame at both ends; this must be jointed true on the top and outside, and notched down on the frame two inches, with a mortise made down through one end at the shoulder, through which a key is driven to keep it firm in its place. This piece of scantling makes the rest, as well as the gauge or guide for the tool; the tool used is either a large jointer plane-iron, or a tool of similar form made for the purpose; instead of the iron caps being screwed on above the plane-iron, a wooden guide is screwed on crossways below; this guide piece is adjusted to the back edge of the rest, like the stock of a T square, and at such a distance on the iron as to allow the cutting edge to reach the cylinder. If now the guiding corner of the rest be set parallel with the axis of the cylinder, and the cylinder be set in motion, it may be turned true and straight from end to end by moving the cutting tool carefully along the guiding rest.

The emery cylinders for grinding the cards may be ten or twelve inches in diameter, and the full length of the card cylinders; they should be made of dry pine, upon an iron shaft, with room for a driving pulley on one end; they are turned true by the process just described. To put on the emery coating, the cylinder should be heated to preserve the glue in a soft state until the emery can be applied; they are covered with a coat of liquid glue, and the emery sifted on equally all round; it should be pressed into the glue by rolling on a smooth surface, and when sufficiently dry, another similar coat added, until the covering is thick enough. Care must be taken to put on the coating equally, as the cylinder must be true when finished, and this is attained by adding more glue and emery on the low places, which are shown by turning it round against a straight rest. We have succeeded in turning these true by laying a large heated bar of iron behind, and holding a burr block against the high places, but it is a tedious process, and injures the coating. The composition is strengthened and wears better by having some emery flour mixed in with the liquid glue

The cards are ground by running them backwards, and running the emery cylinder swiftly against the points of the teeth. They are sometimes ground against an emery board. After being ground until the teeth are equal and sharpened, there is a little rough point or hook left on the inner corner of the wire that catches the wool, and makes trouble at first; these are like the wire edge on a new-ground tool, and should be brushed off by placing the teeth of two cylinders slightly in contact and running them backward; this must be done carefully and well, and is then a great improvement. Such emery cylinders are frequently used to grind cast-iron cylinders that are too hard, or rods that are too slender to be cut by a turning tool, and this is an easy and expeditious way of truing such.

CHAPTER XX.

WINDMILLS.

Fig. 43.*

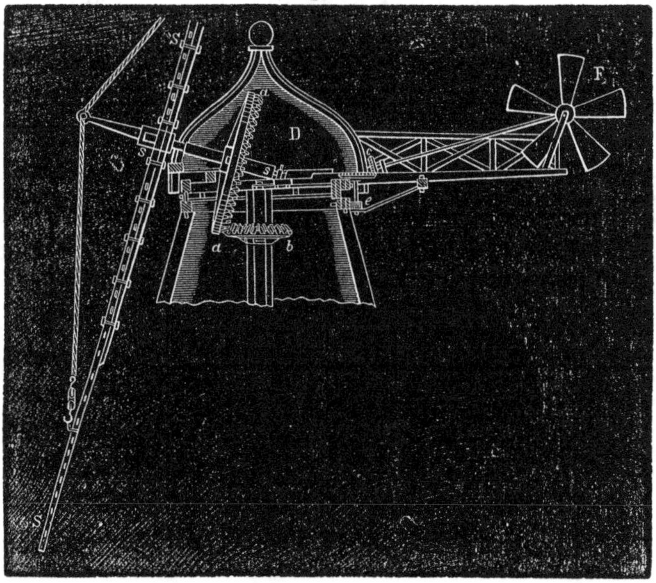

* This mill, from Fairbairn's "Mills and Millwork," is therein described as follows :—

"D is the cap moving on rollers; *s* the shaft carrying the sails, SS and the bevel-wheel *aa*, gearing into another bevel-wheel *b*, on the

We have tried to study out the reason why wind has been almost entirely abandoned in this country as a motive power for propelling machinery, while nearly every other source of power has been improved and brought to such perfection. We remember many years ago, while young, in passing down the St. Lawrence River from Lake Ontario, that windmills were seen busy at work on almost every prominent point; and after getting down into the valley of the St. Lawrence, the windmill was the most picturesque feature in almost every landscape view. Lachine and Laprairie had their windmills, and Montreal not less than half a dozen in a cluster, around the point where the Lachine Canal now disburses its last and greatest water power. That the water-power furnished in such abundance by that canal, should supplant the use of the fickle and less reliable wind on that particular point, is natural enough; but why windmills in other localities where no other power has been substituted instead, should be abandoned to ruin and decay is not so easily understood; and it is still more difficult to find a satisfactory reason for the fact that windmills are so little used on the great prairies of the west, where water can seldom be had, and fuel for the generation of steam is equally scarce. That this oldest and most universal of all motive powers, should be so little employed on these prairies appears still more strange when

mill-stone shaft. The wind acting on the fan F, communicates motion to the bevel-wheel and spur pinion *e*, which, acting on the spur wheel or rack fixed on the summit of the tower, causes the revolution of the cap. The sails of the fan are constructed so that when they lie in the plane of the wind they are not affected; but as the wind shifts, it strikes them obliquely and causes the revolution of the cap till they are again in the plane of the wind."—*Fairbairn's Mills and Millwork*, Part I. pp. 277–8.

we consider that the wind blows over these vast plains in a steady and almost uninterrupted breeze for a great portion of the year without local obstruction, having neither mountain nor hill and scarcely a tree to interrupt or disturb its course.

It is these disturbing influences which abound in the Eastern and Northern States that render the wind too unsteady to be relied upon for driving machinery. These objections will always tend to prevent the use of wind for driving machinery in all extensive manufacturing establishments in the east or north. We have been informed by old country millwrights and millers of extensive experience and observation, that nowhere on the eastern part of this continent does the wind blow sufficiently steady to be reliable for driving extensive machinery. If this is the case, some cause more potent than the local interruptions must interfere to disturb the progress of the wind over the continent. It is said to be an established fact from meteorological observation, that the prevailing winds come from the northwest during three-quarters of the year on an average, over the whole North American Continent. This is the reason given for the average winter temperature being ten degrees colder on the Atlantic than on the Pacific coast in the same latitude. This northwest wind that visits us comes from the northern regions of the Pacific, over the Alaska purchase and the Rocky Mountains. As it passes over the latter obliquely, its course is interrupted and changed, and still more modified and broken by the peaks and passes, and perpetual snow of those mountains. The wind may continue to be affected by these disturbing causes as it passes over the prairie regions and until it arrives among the local obstructions of hills and hollows, forests and fields. We cannot vouch for the entire correctness of this theory, but have always been told by the millers attending the windmills on the St. Lawrence, that the only wind they could rely upon, for running steadily, was that blowing down the river toward the great Gulf. And repeated observations have made us familiar with the effect produced by wind blowing obliquely against a mountain range. Our principal falls of snow come from the east, but the drift seldom takes place until the wind changes to its ordinary point a little north of west. In any level valley where the wind has its free course, the snow drifts follow this direction, but in southwestern New York the wind is turned aside by the Alleghany Mountains, and the drift crosses an east and west road obliquely from the south. The same thing occurs along the whole northern frontier of New York State, where the Adirondac Mountains cause a northern slope that extends from their base beyond the Canada line; and throughout this whole region the deflection of the wind by these mountains is indicated by the direction of the snow drifts, which diverge more or less towards the north according to circumstances; the influence being perceptible far beyond the inclined base of the mountains, upon the level plain. This divergence causes eddies and conflicting currents of wind, which are shifted by the slightest variation of intensity or direction of the breeze, and must tend in a great measure to destroy its useful effect as a motive power. Another effect, still more potent, by which those mountains influence the action of the wind over this northern slope and the contiguous valley, is, that a strong wind from any southern point is thrown upward by the mountain, as a similar volume of water is thrown up by a like sloping obstruction interposed in its chute, the water is soon brought down again by its gravity to the direct course; but it is different with the wind, which has no such tendency to make it resume its original course, and it continues upward, crossing and mingling in with the uninterrupted current above, and thus forming whirlwinds and eddies, with partial vacuums intervening, which reach the slope and the valley, sometimes intensified, and always in the wildest confusion, and frequently at a great distance from the mountains. It is these southern winds, coming over the mountains, that unroof buildings and uproot the trees throughout all this section, and not the uninterrupted winds from any other quarter. A question of the greatest importance to any one contemplating the building of a windmill on any of the western prairies is, how far and to what extent these disturbing influences affect the prevalent westerly winds in their passage over the Rocky Mountains.

We have been consulted as to the expediency of erecting large grist and saw-mills on the western prairies, and explained these theories in detail, the mere outlines of which we have given here, advising the parties to investigate thoroughly before deciding either way. The result was, that after exploring a great portion of the prairie regions, making observations and collecting information from various sources, they gave up the project and returned home. But their investigations were too superficial, and the data upon which they founded their conclusions too vague and unsatisfactory to be reliable; and it still remains in our mind a question, that is yet to be solved by a more competent and careful investigation, to what extent these disturbing phenomena affect the wind as a motive power, and how far their influence extends.

This is an important problem for any person who meditates the building of a windmill anywhere in this country, and one which we are not competent to solve. This much we may say, that although this continent is not so well adapted to windmills as Holland, or some other countries or islands but little elevated above the sea, yet there are no doubt many localities where these would work and pay well. It would require careful and matured observation, with good tact and discrimination, to determine the most suitable site, otherwise as great a disparity would be found between different wind powers, although the machinery were all alike, as is found between different water powers with every variety of machinery.

Thus far we have only contemplated the employment of wind for driving heavy and extensive machinery on a large scale, requiring great and constant power, and involving a corresponding amount of expense, hence the prominence given to the selection of a site where the strongest and the steadiest power might be obtained; and to the effect of outside interferences, that might obstruct the full development of these. We will now consider it the medium by which motion may be given to light machinery, adapted to an almost endless variety of purposes and manipulations, which are generally performed by manual labor or horse power. To many of these operations we have seen these small windmills applied, with little expense and labor; they are especially adapted to farmers and mechanics in country places and small villages, for threshing and winnowing grain, driving a corn-sheller or straw or root-cutter in the barn, to saw wood, pump water, drive the grindstone, churn, or washing machine, at the wood shed, or to drive every description of light machinery in a work-shop, such as lathes, circular, jigger, scroll, and butting saws, bellows, trip hammers, or any other machinery employed by the tradesman that does not require to be run constantly. We have seen nearly all such operations performed by wind. A little experience soon teaches a man the probable and proportionate time the machinery will run, and he learns to calculate his work accordingly. A farmer whose barn and wood shed are furnished with such wind power, can do up all his indoor work in stormy weather, when little can be done to advantage in the open air.

We will endeavor to give such a description in detail as will enable a man of ordinary abilities to construct and set up the wind-wheel required for such a machine as will furnish from one to four horse power in an ordinary stiff breeze, and also instructions for attaching it either by bevel gearing to communicate a revolving motion, or by a crank to produce a reciprocating motion.

Any ordinary building, as a barn, wood-shed, or workshop is strong enough to carry such machinery. In commencing to erect the necessary elevation above the roof upon which to place it, the first thing is to make a hole through the ridge of the roof, as if for a chimney, about three or four feet square, and set up four posts through this hole reaching from the floor or beams below to seven or eight feet above the roof. Round spruce poles five or six inches in diameter make good posts; these should be placed two and a half or three feet apart at the upper end, and secured by caps, framed square on the inside, and circled on the outside. This forms the foundation for the track on which the turn-table with the wind-wheel and its machinery traverses. The lower ends of these posts should be placed about eight feet apart, and framed into sills fastened to the beams or floor on which they rest. One or more sets of girts may be spiked on to these posts at a proper height to lay a floor upon, for convenience, and to strengthen the frame.

There are many different ways of making the wind-wheel, all more or less self-regulating. Some are placed upon a perpendicular shaft, and the vanes revolve horizontally, but the greatest number, and those giving the best results, revolve vertically upon horizontal shafts. The one we prefer, and which admits of the easiest control, is made and regulated as follows:—

The arms are made of two pieces of oak, elm, or ash scantling, fourteen feet long and four inches square; these are halved or notched across each other at the centre, and make the four arms. The arms are left the full size one foot from the centre, and from there they are tapered down to two inches at the end, the tapered part being rounded. Three bearings two inches long, like journals, are made on the round part of each, by leaving a portion of the corners for a shoulder: one eighteen inches from the centre, another on the small end near the point, the third one half way between the others. Cross-bars are fitted on the face of each of these bearings, and secured by a cap screwed on behind, the bar next the centre eighteen inches long, the middle one twenty-four, and that at the point thirty inches long.

The sails, made of thin boards, are nailed upon these cross-bars, which are placed upon the arms so that the sail is not equally balanced, being only nine inches wide from the arm to the weather or foremost edge, while the extra width at the outer ends is all to the rear side of the arm. It will be seen by this arrangement that the sails, if not otherwise controlled, when exposed to the wind, would turn their edges towards it, the long corner being blown to leeward like the tail of the vane, or weathercock; this describes the principle by which this kind of wind-wheel is regulated and rendered safe. The sails thus hung are held more or less obliquely to the wind according to its intensity and the amount of power required, or left free to resume the weathercock position described, and when thus situated they have no power to turn even in a gale of wind. The power to regulate and control the sails is applied through the centre of the main shaft, which is made hollow for this purpose, and an iron rod with a double cross head, which makes four projecting arms, is poked through the hollow until the cross arms are immediately in front of the sails; each of these arms is connected by a jointed rod or link to the rear corner of each sail. The central rod projects through the other end of the main shaft, and is pivoted to the end of a bent steelyard lever. A weight hung upon this lever, like that on a safety valve, graduates the power of the wind and the work as required, and is removed to stop the machine.

An old friction rod, or guide for a saw gate, makes a good shaft for a wind-wheel of this size and construction. A thick disk of plank should be pinned on to the back side of the arms at the centre to strengthen and give sufficient bearing, as the arms should be left whole, except the small hole through which the regulating rod passes. Then an iron flange should be bolted on behind the wooden disk to fasten the whole together when keyed to the shaft. A bevel wheel is placed on this shaft so as to gear into and traverse around a pinion of smaller size on the top of the upright shaft. The upper bearing of the upright shaft is in the centre of the cross-bar of the circular cap which supports the turn-table, and close below the pinion. A band of iron two inches wide and three-eighths of an inch thick is put around this cap, standing nearly half its width above the wood; this forms the track on which the table carrying the wind-wheel traverses. This table is set upon four rollers or small cast-iron wheels with grooves cut in them to fit and hold upon the track. A clamp attached to the table at each wheel, for safety, passes down outside the track, and hooks loosely under the circular cap. A large vane or tail of thin boards is attached to the turn-table opposite the wind-wheel, to keep the latter fair to the wind, and is called a director. The bent lever alluded to, for regulating the sails to the wind, has a pin through the corner for its fulcrum, a socket at the short perpendicular end, for the end of the regulating rod to turn in, and a weight hung on the long horizontal end, to be graduated to suit the wind and work. The motion can be taken from any part of the central upright shaft by a belt. If for a horizontal motion, the belt must be put on with a half twist, and the slack side brought to the proper line by a pulley set at an angle, the driving side being straight (see the article on belt grist-mills). Or a horizontal motion may be taken by bevel gearing, if required or preferred. A reciprocating or oscillating motion may be taken from any of these revolving motions by a crank.

A crank is sometimes made in the centre, on the wind-wheel shaft, in place of the bevel wheels, and the motion transmitted down through a pitman and connecting rod, jointed together by a swivel; but the motion is too slow for ordinary purposes, and we have never seen this plan used except for pumping. When the power has to be transmitted to a greater distance than shafting, belts, or chains admit of, it can be easily and cheaply done through the medium of wire. If a reciprocating, for sawing, churning, pumping or the like, it may be transmitted from one oscillating lever (walking beam) to another, by connecting both ends of the two levers with wire, and giving out the motion as shown.

If a revolving motion is to be transmitted, a continuous wire is used, and passed around two light wheels, but large in diameter, the wire in this case driving like a belt. Sometimes the wire is passed clear round the wheels in a groove, sometimes only half round, like a belt, and the groove edge of a tightener pressing it against the back of each wheel to give it sufficient bite; but the best way is to cover the edge of both wheels with a strip of sole leather or Indian rubber, and the wire will hold sufficiently either cross or open banded, the weight of the stretches of wire suspended between giving sufficient friction and elasticity.

By these wire connections a light and easy movement may be carried to a distance of several hundred feet, but when a single wire is used, the size of the wheels around which it works must be increased as the thickness of the wire is increased, because the bending of the wire must not be so acute as to permanently affect its elasticity. And when a considerable amount of power and velocity is thus transmitted, a wire rope made of several strands of small wire must be used. In all cases where the distance is very great, the slack of the wire must be held up by intermediate pulleys.

By erecting a suitable wind-wheel on this principle, in a central situation, with a judicious distribution of its power by the means indicated, to the different buildings, all the operations about a farmer's premises may be performed in their turn by the same wheel, and a great amount of hard labor saved.

This is a power that every person owning a farm or building possesses, although some localities are better situated than others for obtaining a steady wind power, and we firmly believe if it were less common, and occurred only here and there at wide intervals like water power, it would be more valued and more used.

Sails made like those described are sometimes regulated by heavy cast-iron balls hung to the outer end of each arm by bent levers, which turn the face of the sails away from the wind by centrifugal force. Other sails are made of small pieces of thin boards hung in a frame similar to window blinds. These blinds are connected together at the edges by strips running the whole length of the arm, and all moved at once by a centrifugal governor driven by the machinery inside.

The oldest method we have seen is a wooden rack with a canvas sail spread upon it, the canvas being more or less twisted up or spread out at the inner end to vary the amount of surface exposed, according to the wind.

These are sometimes furled or twisted up by a governor driven by the machinery, but more frequently this is done by hand. It will be observed that none of these methods of regulating have any control over the sails, except when under full motion, and are therefore of no use in stopping the machine or saving it in a gale when standing idle.

Sometimes the lath sails alluded to become so encrusted with ice, in a cold storm of rain and sleet, that the hundred journals upon which the slats turn are frozen firmly in their sockets, with no means of stopping the mill but by applying the brake, and a general breakdown is often the consequence. Such a condition of things prevents the furling of the canvas sails, and produces a similar result.

If the sails of a windmill were placed with their faces perpendicularly to the wind, it would have no power to turn them either way, therefore the sails must be set at a certain angle. This is called the weathering of the sails, and the degree of inclination should vary at different distances from the centre. It is found in practice that the angle need not be varied much for the first half from the centre, on account of the interruption which the wind meets with from the hub and framework, and the close proximity of the different sails. An angle of about eighteen degrees with the plane of motion will answer well for the inner half. From the centre of the sail the angle should begin to diminish until not more than eight degrees at the extreme outer end. This furnishes a clue to the reason why such a wind-wheel revolves so much faster and stronger than one of equal area of sails will revolve horizontally. The vertical wind-wheel is held and revolves in the same plane, with the inclined face of every sail constantly exposed to the force of the wind throughout the whole circuit, and the velocity of its motion is thus screwed up beyond the velocity of the wind that drives it, and the number of its revolutions is increased in proportion to the obliquity of the inclined face of the sails. When the horizontal wind-wheel is compared with this kind, its disadvantages are apparent; in the first place the wind can only act upon one side, and that side receding away from the wind, while the opposite side must move directly against it; then the velocity with which the working sails move must be deducted from the velocity of the wind, and the balance is all the velocity with which the wind can impinge upon the sails. And as the sails recede all on the same level, and near the same line, those behind intercept and break the force of the wind before it strikes the sails in advance, and thus instead of intercepting the full force of a cylinder of wind equal to the whole circumference and area of the wheel like the other, the horizontal wheel only intercepts a portion of the area of one side, and even that side in a confused and broken manner. Add to all these disadvantages, the continual clanking and cracking of the sails as each one changes its face and edge alternately to the wind, in passing the two centres, also their liability to get out of order, and the reason will be sufficiently obvious why every experienced millwright condemns this method of applying the wind to drive machinery, and the horizontal wind-wheel is only used by amateurs or visionaries.

Perhaps we might be able to convey a more intelligible idea of the philosophy of the weathering or oblique action of the vertical sails, by comparing these to similar contrivances embracing the same principle, both in nature and art; the wings of a bird and the fins and tail of a fish exhibit the principle to perfection, but these being equally balanced, as attached to the fowl and fish, the identity is less apparent. A better illustration of the principle is shown by the revolutions of the seeds of the maple, and some other forest trees, in descending from the branches to the ground; each seed is furnished with a wing or fin (sail) resembling the wing of a dragon-fly, and as the seeds which grow in pairs break apart when ripe, and fall singly, the wing or sail revolves rapidly around the seed as a centre, and thus sails to a great distance from the tree before touching the ground. This is a mechanical provision, by which nature assists the wind to scatter those seeds over the earth, and no doubt furnished the aborigines of Australia with the idea which they developed in the boomerang.

The rapid facility of revolution which these winged seeds and the boomerang exhibit in their progress through the air, probably originated the idea of navigating and steering a balloon by means of similar revolving wings or sails, propelled by some light and strong motive power; and this idea has a better prospect of being successfully employed to accomplish that purpose than any other yet proposed.

Perhaps a still better illustration of the power which the wind possesses to impart a rapid motion when applied obliquely to the plane of that motion may be seen in its action upon the sails of a vessel on the water, or an ice boat on the ice. When the vessel or boat is moving in the same direction as the wind, its sails are in the same predicament as those of the horizontal windmill, that is, they are moving away from the wind, and their velocity is deducted from that of the wind as it impinges upon them, and the rear sails obstruct the wind and lessen its effect upon those in advance; but when the vessel or boat is moving in a line obliquely to the direction of the wind, the sails are situated like those of the vertical windmill, that is, they are held to the wind by the keel of the vessel, and the runners of the ice boat, as the sails of this kind of windmill are held by the shaft, and thus get the full unobstructed force of the wind upon each sail, with the increased velocity due to the obliquity of their position.

CHAPTER XXI.

STEAM POWER.

THE application of steam in propelling machinery has been so much improved, and its production and management have been brought under such control, that it is rapidly approaching a degree of perfection that has already enabled it to supplant, in a certain degree, all other motors, whether of wind, water, or animals. This, with its adaptation to all situations where water and fuel can be had, and its unlimited range as to the amount of power, will enable it to supersede these old motive powers still more in the future than it does at the present time.

The wind, although it costs nothing, and is so universally distributed, is proverbially unsteady, and therefore not sufficiently reliable to drive the machinery for this progressive age and nation. Water powers, although the steadiest of all powers, occur only in a comparatively few localities, and these are often in remote, and frequently almost inaccessible situations, far from the great centres of population, and therefore the greatest and best water powers always have been, and possibly always will be, unoccupied. We have little to say of the comparative convenience or cost of animal power, because the evident repugnance of the animals as they toil and sweat to give motion to machinery, will always tend to prevent their general employment, where inanimate forces can be used.

With regard to the comparative *cost* of any given amount of power to be furnished by these different motors, it is not possible to find any data that would be generally applicable for even an approximate estimate. Each particular situation, with all the circumstances and contingencies affecting it, must form an item in such calculations.

The wind costs nothing, whatever amount may be used, but a suitable wind-wheel and machinery to employ it are expensive; then a building sufficiently high and substantial upon which to secure it in the necessarily elevated situation, with the vanes and machinery required to keep it right to the wind—these are items of first cost; then the expense of running and keeping it in repair. Against these expenses, the profits on the probable amount of work performed are to be placed, making due allowance for delays and the expense and loss when thus idle.

It is still more difficult to estimate the actual average cost of a given amount of water power; the range of all the contingencies in each particular site being so much more varied. There is the dam, an important item; then the excavation for the foundation and mill-race; the building with its yard and approaches—these are frequently very expensive items; and lastly, the water-wheel and necessary machinery. These, when accurately found, and all added together, will show the first cost of the water power. To this must be added the probable expense of working and repairs. With this aggregate, the profits estimated, as in the other case, with due regard to contingencies, must be balanced and compared.

The first cost and working expenses, as well as all the ordinary contingencies affecting the operation of a steam engine, of a power equal to these wind and water powers, can be more easily and accurately calculated, because the steam engine is now reduced to fixed rules of weight and measure by which its cost and power may be relatively estimated. Its expenses in working, although much greater than either of the others, can also be more accurately figured up, as the contingencies are subject to human control in the engine, being nothing more than generating the proper supply of steam.

In comparing and deciding upon the most suitable of these powers for a particular business, in a given locality, all the various items of expense, and all the contingencies affecting each, must be carefully noted *as they occur in that locality;* and when this is done by a competent person, having reliable data, it will be found that wind is the most economical for a certain business and location, water for another, and steam will be decided upon as the most suitable agent as often as both the others. With regard to the liability to accidents, the three may be put on a par; the probability and effect of a hurricane on a windmill, a freshet on a water-mill, and an explosion on a steam-mill being about equally balanced.

In general terms, it may be said that where water power can be had to do the required work, it is to be preferred before all others. Where water power cannot be had without going into inconvenient situations, and fuel to generate steam is plenty, then a steam-engine is preferable, as the free selection of the site is often of much more value than the fuel. For instance, a person will go back into an out-of-the-way ravine and build a saw-mill, where the clearing away and burning of the slabs and edgings will cost as much as the firing of an engine. Then all his lumber must be hauled out to market, and his supplies hauled in, his logs costing perhaps as much *there* as they would cost at a village or depot, or navigable point of a river, where the slabs and edgings would bring an annual revenue, which, with the cost of hauling saved, would make a handsome profit. Experience shows that the sawdust and waste bark are more than sufficient for the firing; and the first cost of the steam-mill would generally be less than the other.

When the work has to be performed in the absence of water power, and where fuel for the generation of steam is too scarce and dear, then the clearest and most elevated situation should be selected, and a windmill built of sufficient capacity. This will furnish a large amount of power in the course of a year, at little expense; but calculations must be be made to "bide its time," and something provided to employ the hands while thus waiting.

The various methods of applying water and wind for propelling machinery have already been mentioned; the steam-engine furnishes the method of applying steam to that purpose. But we do not propose to give any particular description of this motor, because its details are much better described by able engineers than we could expect to give them; and the millwright has little business with it further than to set it in proper position and connect it, as he would a water or wind wheel, with the machinery to be driven.

There are many different kinds of engines, and the makers generally furnish directions more to the purpose than we could give for setting their particular engine up and working it. We would therefore mention, as the most important point to be determined, after deciding to use steam, which of all the various kinds of engines will best suit the situation and circumstances? If fuel be expensive and room valuable, then a flue or tubular boiler embracing more or less of the locomotive principle should be used. We have seen a partly worn railway locomotive boiler, considered unsafe for the road, put to a stationary use, and answer a good purpose for many years. For driving a saw-mill or other wood working machinery that will furnish its fuel in sawdust or other refuse, or in the woods or coal regions, where fuel is cheap as well as room, a less delicate and more substantial form of boiler and engine, with a more capacious furnace and grate surface, will be more suitable.

There is such a wide difference in the plan and construction of the various engines, and boilers, and furnaces, and also in the kind of work and situation in which they are to be employed, that the selection of the kind most suitable in every respect is an important question, which should never be decided by mere accident, as it frequently is; but the assistance of a competent person and reliable information should be had to decide this point.

It may be said to be established by experience that steam cannot compete successfully with water power on *equal terms*, for driving heavy machinery, even where fuel is cheap. But the facility which steam power gives of selecting the most suitable situation for carrying on a particular business gives it a decided advantage over water power, especially for light machinery. A man in a city or village can set up a small engine, and thus enlarge his former business, or start a new one, with little or no expense for new buildings. A heavy business, as a grist-mill, may also be established where the saving of transhipment and hauling, added to the better market and enhanced value of the bran and offal, would give a good margin for profit in favor of steam over the nearest water power.

APPENDIX.

MERCHANT BOLT.

CONSTRUCTED BY HENRY SMITH, JR., MILWAUKEE, WISCONSIN.

DESCRIPTION OF PLATES.

FIG. 1.—*Showing Machinery Posts and Pitch Lines of Wheels.*

A. Posts and caps, 6 x 10 inches.

B. Bridge trees, with keys.

C. Spur wheels of 140 teeth, 1 inch pitch, 2¼ inches face, 3 feet 8 inches and $\frac{54}{100}$ diameter.

a a. To drive upper reels for flour.

b b. To drive lower reels, or return reels.

c c. Main driving wheel; on same shaft is a spur wheel, *D*, of 114 teeth, 1 inch pitch, 2 inches face, 3 feet $\frac{48}{100}$ inch diameter, which connects with a spur wheel marked *E*, of 50 teeth, 1 inch pitch, 2 inches face, which drives flour conveyor, being fastened in wooden conveyor shaft by a gudgeon. (See figs. 5, 6, 7.)

F. Intermediate spur wheels of 60 teeth, 1 inch pitch, 2 inches face, 1 foot 7$\frac{20}{100}$ inch diameter.

G. Spur wheels of 60 teeth, 1 inch pitch, 2 inches face, driving return conveyor. On same shaft is a spur wheel marked *H*, which has 50 teeth, 1 inch pitch, 2 inches face, 1 foot 4 inches diameter, which connects to a spur marked *H H*, which is attached to cut-off conveyor.

J. Floor place on joist marked *K.*

L. Beams, 12 x 12 inches square.

d d. Spur wheels driving dusting reels, which are driven by spur wheels marked *N*, of 96 teeth, 1 inch pitch, 2¼ inches face, 2 feet $6\frac{72}{100}$ inches diameter. This spur wheel is driven by an upright shaft running in front of chest, and connects with a pair of bevels at the shaft that drives wheel marked *cc* in upper chest. See fig. 5, *KK*, and *LL*.

M. Spur wheels of 82 teeth, 1 inch pitch, 2 inches face, 2 feet $2\frac{24}{100}$ inches diameter. These two wheels drive conveyor for dustings.

FIG. 2.—*Showing the end of Bolt Chest next to Machinery Posts, which will be seen by comparing the Bridge trees, except the lower one where the shaft rests in the Conveyor Case, by means of a Box fitted into Conveyor end; this end of chest is the front.*

A. Posts and caps 3 x 12, with bridge trees marked *B*.

C. This is the discharge end of upper reels or flour reels.

E. This is a board partition around flour reel, giving the reel ½ inch play; all that is discharged from flour reels runs down spouts *E*, into the return reels marked *D*.

F. Are the cant boards leading flour in flour conveyor marked *G*. *gg* are 2 x 4 pieces put underneath cants *F* to stiffen and support them.

H. Is the support of flour conveyor, is a 2 inch plank made to rest on cant boards *J*.

J. Are lower cant boards, supported in the centre by means of a 4 x 6 inches marked *K*, bevelled on upper edge and resting on posts of 4 x 4 inches marked *L*.

M. Return conveyor showing how cut-off spouts are constructed. These cut-off spouts are placed at a distance of 10 inches, and are 7 inches in the clear. This conveyor is for the purpose of grading returns.

N. Cut-off conveyor receives all that is cut off from return conveyor and the middlings, also partly conveyor returns as seen in the side view fig. 5, and discharged by spouts *bb* into main meal elevator, and returns to cooler.

X. Discharge spout from upper reel, and also feed spout for return reel *D.* Compare *E* in fig. 2.

Small fig. 1 in side view are the two spur wheels marked *a a* in fig. 1. 2 is spur wheel marked *c c* in fig. 2. 3 is spur wheel *bb* in fig. 1. 4 compare with *F* in fig. 1. 5 with *G* in fig. 1. 6 with dotted lines *D* in fig. 1. 7 with dotted lines *E.* 8 with dotted lines *H* in fig. 1. 9 with *H H* in fig. 1.

b b. In fig. 5 is a conveyor that receives the flour from chest by spout *SS*, and conveys the same into packer chest.

Red lines in fig. 1 show distances in feet and inches, the same in fig. 5 show the centre of shafts.

The arrows in fig. 5 show which way conveyors carry.

The bridge trees are made of pine, 9 inches wide and 3 inches thick, the boxes are let into the same, and are 4 inches long, with an ½ inch tenon on each end.

Cap let into bridge-tree, ¼ inch deep.

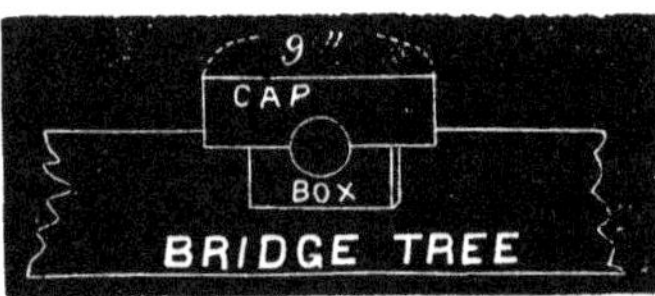

O. In fig. 2 are the beams in mill building, 12 x 12.

P. Dusting reels for bran and middlings. *P* 1 is bran duster, on tail end of same is 2 feet of wire cloth to remove dough balls, &c. &c., and prevent the same from entering bran duster. They are discharged from chest by spout *Y.* The bran after it leaves reel drops in spout *U*, to bran duster *W.*

P 2, is middling duster, the dusted middlings drop from reel into spout *R*, to elevator *S*, to be carried off the middling stone.

Q. Forms the discharge spout around reels or partition, as seen in side view fig. 5, marked *g g.*

T. Dusting conveyor, conveys all dustings to main meal elevator and returns to cooler by spout *T* 2, the dustings from conveyor marked *T*2, drop into a spout and meet spout *T* 2.

X. Girts to support conveyors, &c. &c. (See side view, *P P*.)
Z. Partition between the reels made of one inch lumber, matched.
V. 4 x 6 whole length of chest to support cant boards, &c. &c.

FIG. 3.—*Showing Back of Chest.*

A. Posts and caps in frame.
B. Bridge tree, 6 inches higher than the bridge tree on front of chest.
D. Flour reels, showing the head where flour enters from cooler by spouts *G*. *E*, forms speck box around reels, giving reels ½ inch play, and made of 1 inch lumber, matched. (See fig. 5, *a a*.) This catches specks and drops them in the return conveyor by spout *E*. (See also *W*, in fig. 5.)
F. Cant boards, 1 inch lumber matched, stiffened by 2 x 4, marked *d d*.
G. Flour conveyor discharge by spout *H*, on fig. 5. (See spout *s s*.)
J. Return reels lower end, or discharge end. The bran comes out of tail end, and is collected and dropped by partition *K*, and spout *K*, from two reels in one dusting marked *R*.
L. Bridge tree, 6 inches lower than at head.
M. Cant boards. *N*. Return conveyor. *O*. Middlings conveyor.
P. Middling spouts to middling duster. *P* 2 can be changed so as to come under conveyor, or to receive from conveyor *O*.
R. Showing bran duster reel head, by circle lines. *P* 1, in fig. 2, is same reel.
S. Middling duster reel. Conforms with *P* 2, in fig. 2.
T. Bridge tree, 6 inches higher than same in fig. 2.
V. Beam in mill.
W. Partition between reels.
X. Cant boards, supported in centre by means of a 4 x 6 marked *Y*.

Z. Dusting conveyors. (See *T* 1, and *T* 2, fig. 2.)
a a. Girt, same as *X* in fig. 2.

FIG. 4.—*Ground Plan, showing the placing of Posts of Frame.*

A. Corner posts on chest, 3 x 12. (See side view *A*.)
B. 3 x 6 Posts. (See fig. 5, small post marked *A*.)
C. Machinery post, 6 x 10. (See fig. 1 *A*.)

FIG. 5.—*Side view of Chest.*

A. Posts of frame.
B. Bridge trees, end view.
C. Upper or flour reels.
D. Return reels.
E. Dusting reels.
F. Flour conveyor with drop slides, as seen at *G*; they are handled by small lines running over sheaves above every slide, and running out of chest at back end.
H. Return conveyor with cut-off slides, seen at *h h*, let in ½ in bottom of conveyor so as to make slide ⅝ of an inch thick. These slides are made to cut off returns if too coarse, and put into middlings by dropping into conveyor below marked *J*, which carries the same into middling duster reel by spout *Z*. Compare with *P*, in fig. 3.
K. Return spout, carrying returns to meal elevator. This spout runs through floor and along side lower reels and drops in conveyor *N*, and from conveyor *N* by spout *Y*. (See *T* 2, fig. 2.) Also back of spout *U* in fig. 2, there is the same spout which runs into spout *T* 2, in fig. 2, next to floor.
L. Floor of mills laid double of 1 inch boards.
M. Joist 3 x 12.
O. Girt extending the whole length of chest, supporting cross girts *P P*; this long girt is supported by small posts marked *A* 1, which are fastened to joists above,

the whole of which carries the conveyors and cant boards.

P. Beams in mill, 12 x 12.

R. Bridge pot, or step for upright shaft, marked *R R*, by red line, which upright is driven by bevel wheel *K K*, on bevel pinion *L L*, connecting on lower chest by *N N*, on *M M*.

K K. Is a bevel wheel of 70 teeth, 1⅜ inch pitch, 3¾ inches face.

L L. Is a bevel pinion to match, 60 teeth, 1⅜ inch pitch, 3¾ inches face.

N N. Is a bevel wheel of 65 teeth, 1¼ inch pitch, 3¼ inches face.

M M. Is a bevel pinion of 52 teeth, 1¼ inch pitch, 3¼ inches face.

S. Reel shafts. Shafts made out of 8 x 8, 22 feet long, six square, which is at the option of millwright.

T. Feed spout from cooler; bolt is through this spout by means of shoe or Cornwell's feeder.

W. Speck spout. Compare with small *e* in fig. 3.

INDEX.

CATALOGUE
OF
PRACTICAL AND SCIENTIFIC BOOKS,
PUBLISHED BY
HENRY CAREY BAIRD,
INDUSTRIAL PUBLISHER,
No. 406 WALNUT STREET,
PHILADELPHIA.

☞ Any of the Books comprised in this Catalogue will be sent by mail, free of postage, at the publication price.

☞ This Catalogue will be sent, free of postage, to any one who will furnish the publisher with his address.

ARMENGAUD, AMOUROUX, AND JOHNSON.—THE PRACTICAL DRAUGHTSMAN'S BOOK OF INDUSTRIAL DESIGN, AND MACHINIST'S AND ENGINEER'S DRAWING COMPANION: Forming a complete course of Mechanical Engineering and Architectural Drawing. From the French of M. Armengaud the elder, Prof. of Design in the Conservatoire of Arts and Industry, Paris, and MM. Armengaud the younger and Amouroux, Civil Engineers. Rewritten and arranged, with additional matter and plates, selections from and examples of the most useful and generally employed mechanism of the day. By WILLIAM JOHNSON, Assoc. Inst. C. E., Editor of "The Practical Mechanic's Journal." Illustrated by 50 folio steel plates and 50 wood-cuts. A new edition, 4to. . $10 00

ARROWSMITH.—PAPER-HANGER'S COMPANION: A Treatise in which the Practical Operations of the Trade are Systematically laid down: with Copious Directions Preparatory to Papering; Preventives against the Effect of Damp on Walls; the Various Cements and Pastes adapted to the Several Purposes of the Trade; Observations and Directions for the Panelling and Ornamenting of Rooms, &c. By JAMES ARROWSMITH, Author of "Analysis of Drapery," &c. 12mo., cloth $1 25

BAIRD.—THE AMERICAN COTTON SPINNER, AND MANAGER'S AND CARDER'S GUIDE:

A Practical Treatise on Cotton Spinning; giving the Dimensions and Speed of Machinery, Draught and Twist Calculations, etc.; with notices of recent Improvements: together with Rules and Examples for making changes in the sizes and numbers of Roving and Yarn. Compiled from the papers of the late ROBERT H. BAIRD. 12mo. $1 50

BAKER.—LONG-SPAN RAILWAY BRIDGES:

Comprising Investigations of the Comparative Theoretical and Practical Advantages of the various Adopted or Proposed Type Systems of Construction; with numerous Formulæ and Tables. By B. Baker. 12mo. $2 00

BAKEWELL.—A MANUAL OF ELECTRICITY—PRACTICAL AND THEORETICAL:

By F. C. BAKEWELL, Inventor of the Copying Telegraph. Second Edition. Revised and enlarged. Illustrated by numerous engravings. 12mo. Cloth $2 00

BEANS.—A TREATISE ON RAILROAD CURVES AND THE LOCATION OF RAILROADS:

By E. W. BEANS, C. E. 12mo. (In press.)

BLENKARN.—PRACTICAL SPECIFICATIONS OF WORKS EXECUTED IN ARCHITECTURE, CIVIL AND MECHANICAL ENGINEERING, AND IN ROAD MAKING AND SEWERING:

To which are added a series of practically useful Agreements and Reports. By JOHN BLENKARN. Illustrated by fifteen large folding plates. 8vo. $9 00

BLINN.—A PRACTICAL WORKSHOP COMPANION FOR TIN, SHEET-IRON, AND COPPER-PLATE WORKERS:

Containing Rules for Describing various kinds of Patterns used by Tin, Sheet-iron, and Copper-plate Workers; Practical Geometry; Mensuration of Surfaces and Solids; Tables of the Weight of Metals, Lead Pipe, etc.; Tables of Areas and Circumferences of Circles; Japans, Varnishes, Lackers, Cements, Compositions, etc. etc. By LEROY J. BLINN, Master Mechanic. With over One Hundred Illustrations. 12mo. $2 50

BOOTH.—MARBLE WORKER'S MANUAL:

Containing Practical Information respecting Marbles in general, their Cutting, Working, and Polishing; Veneering of Marble; Mosaics; Composition and Use of Artificial Marble, Stuccos, Cements, Receipts, Secrets, etc. etc. Translated from the French by M. L. BOOTH. With an Appendix concerning American Marbles. 12mo., cloth . . $1 50

BOOTH AND MORFIT.—THE ENCYCLOPEDIA OF CHEMISTRY, PRACTICAL AND THEORETICAL:

Embracing its application to the Arts, Metallurgy, Mineralogy, Geology, Medicine, and Pharmacy. By JAMES C. BOOTH, Melter and Refiner in the United States Mint, Professor of Applied Chemistry in the Franklin Institute, etc., assisted by CAMPBELL MORFIT, author of "Chemical Manipulations," etc. Seventh edition. Complete in one volume, royal 8vo., 978 pages, with numerous wood-cuts and other illustrations. $5 00

BOWDITCH.—ANALYSIS, TECHNICAL VALUATION, PURIFICATION, AND USE OF COAL GAS:

By Rev. W. R. BOWDITCH. Illustrated with wood engravings. 8vo. $6 50

BOX.—PRACTICAL HYDRAULICS:

A Series of Rules and Tables for the use of Engineers, etc. By THOMAS BOX. 12mo. $2 00

BUCKMASTER.—THE ELEMENTS OF MECHANICAL PHYSICS:

By J. C. BUCKMASTER, late Student in the Government School of Mines; Certified Teacher of Science by the Department of Science and Art; Examiner in Chemistry and Physics in the Royal College of Preceptors; and late Lecturer in Chemistry and Physics of the Royal Polytechnic Institute. Illustrated with numerous engravings. In one vol. 12mo. . $1 50

BULLOCK.—THE AMERICAN COTTAGE BUILDER:

A Series of Designs, Plans, and Specifications, from $200 to to $20,000 for Homes for the People; together with Warming, Ventilation, Drainage, Painting, and Landscape Gardening. By JOHN BULLOCK, Architect, Civil Engineer, Mechanician, and Editor of "The Rudiments of Architecture and Building," etc. Illustrated by 75 engravings. In one vol. 8vo. $3 50

BULLOCK.—THE RUDIMENTS OF ARCHITECTURE AND BUILDING:

For the use of Architects, Builders, Draughtsmen, Machinists, Engineers, and Mechanics. Edited by JOHN BULLOCK, author of "The American Cottage Builder." Illustrated by 250 engravings. In one volume 8vo. . . . $3 50

BURGH.—PRACTICAL ILLUSTRATIONS OF LAND AND MARINE ENGINES:

Showing in detail the Modern Improvements of High and Low Pressure, Surface Condensation, and Super-heating, together with Land and Marine Boilers. By N. P. BURGH, Engineer. Illustrated by twenty plates, double elephant folio, with text. $21 00

BURGH.—PRACTICAL RULES FOR THE PROPORTIONS OF MODERN ENGINES AND BOILERS FOR LAND AND MARINE PURPOSES.

By N. P. BURGH, Engineer. 12mo. . . . $2 00

BURGH.—THE SLIDE-VALVE PRACTICALLY CONSIDERED:

By N. P. BURGH, author of "A Treatise on Sugar Machinery," "Practical Illustrations of Land and Marine Engines," "A Pocket-Book of Practical Rules for Designing Land and Marine Engines, Boilers," etc. etc. etc. Completely illustrated. 12mo. $2 00

BYRN.—THE COMPLETE PRACTICAL BREWER:

Or, Plain, Accurate, and Thorough Instructions in the Art of Brewing Beer, Ale, Porter, including the Process of making Bavarian Beer, all the Small Beers, such as Root-beer, Ginger-pop, Sarsaparilla-beer, Mead, Spruce beer, etc. etc. Adapted to the use of Public Brewers and Private Families. By M. LA FAYETTE BYRN, M. D. With illustrations. 12mo. $1 25

BYRN.—THE COMPLETE PRACTICAL DISTILLER:

Comprising the most perfect and exact Theoretical and Practical Description of the Art of Distillation and Rectification; including all of the most recent improvements in distilling apparatus; instructions for preparing spirits from the numerous vegetables, fruits, etc.; directions for the distillation and preparation of all kinds of brandies and other spirits, spirituous and other compounds, etc. etc.; all of which is so simplified that it is adapted not only to the use of extensive distillers, but for every farmer, or others who may wish to engage in the art of distilling By M. LA FAYETTE BYRN, M. D. With numerous engravings. In one volume, 12mo. $1 50

BYRNE.—POCKET BOOK FOR RAILROAD AND CIVIL ENGINEERS:

Containing New, Exact, and Concise Methods for Laying out Railroad Curves, Switches, Frog Angles and Crossings; the Staking out of work; Levelling; the Calculation of Cuttings; Embankments; Earth-work, etc. By OLIVER BYRNE. Illustrated, 18mo., full bound $1 75

BYRNE.—THE HANDBOOK FOR THE ARTISAN, MECHANIC, AND ENGINEER:

By OLIVER BYRNE. Illustrated by 185 Wood Engravings. 8vo. $5 00

BYRNE.—THE ESSENTIAL ELEMENTS OF PRACTICAL MECHANICS:

For Engineering Students, based on the Principle of Work. By OLIVER BYRNE. Illustrated by Numerous Wood Engravings, 12mo. $3 63

BYRNE.—THE PRACTICAL METAL-WORKER'S ASSISTANT:

Comprising Metallurgic Chemistry; the Arts of Working all Metals and Alloys; Forging of Iron and Steel; Hardening and Tempering; Melting and Mixing; Casting and Founding; Works in Sheet Metal; the Processes Dependent on the Ductility of the Metals; Soldering; and the most Improved Processes and Tools employed by Metal-Workers. With the Application of the Art of Electro-Metallurgy to Manufacturing Processes; collected from Original Sources, and from the Works of Holtzapffel, Bergeron, Leupold, Plumier, Napier, and others. By OLIVER BYRNE. A New, Revised, and improved Edition, with Additions by John Scoffern, M. B , William Clay, Wm. Fairbairn, F. R. S., and James Napier. With Five Hundred and Ninety-two Engravings; Illustrating every Branch of the Subject. In one volume, 8vo. 652 pages . $7 00

BYRNE.—THE PRACTICAL MODEL CALCULATOR:

For the Engineer, Mechanic, Manufacturer of Engine Work, Naval Architect, Miner, and Millwright. By OLIVER BYRNE. 1 volume, 8vo., nearly 600 pages $4 50

BEMROSE.—MANUAL OF WOOD CARVING: With Practical Illustrations for Learners of the Art, and Original and Selected designs. By WILLIAM BEMROSE, Jr. With an Introduction by LLEWELLYN JEWITT, F. S. A., etc. With 128 Illustrations. 4to., cloth $3 00

BAIRD.—PROTECTION OF HOME LABOR AND HOME PRODUCTIONS NECESSARY TO THE PROSPERITY OF THE AMERICAN FARMER:

By HENRY CAREY BAIRD. 8vo., paper 10

BAIRD.—STANDARD WAGES COMPUTING TABLES:

An Improvement in all former Methods of Computation, so arranged that wages for days, hours, or fractions of hours, at a specified rate per day or hour, may be ascertained at a glance. By T. SPANGLER BAIRD. Oblong folio $5 00

BISHOP.—A HISTORY OF AMERICAN MANUFACTURES:

From 1608 to 1866: exhibiting the Origin and Growth of the Principal Mechanic Arts and Manufactures, from the Earliest Colonial Period to the Present Time; with a Notice of the Important Inventions, Tariffs, and the Results of each Decennial Census. By J. LEANDER BISHOP, M. D.; to which are added Notes on the Principal Manufacturing Centres and Remarkable Manufactories. By EDWARD YOUNG and EDWIN T. FREEDLEY. In three vols. 8vo. $10 00

BOX.—A PRACTICAL TREATISE ON HEAT AS APPLIED TO THE USEFUL ARTS:

For the use of Engineers, Architects, etc. By THOMAS BOX, author of "Practical Hydraulics." Illustrated by 14 plates, containing 114 figures. 12mo. $4 25

CABINET MAKER'S ALBUM OF FURNITURE:

Comprising a Collection of Designs for the Newest and Most Elegant Styles of Furniture. Illustrated by Forty-eight Large and Beautifully Engraved Plates. In one volume, oblong $5 00

CHAPMAN.—A TREATISE ON ROPE-MAKING:

As practised in private and public Rope-yards, with a Description of the Manufacture, Rules, Tables of Weights, etc., adapted to the Trade; Shipping, Mining, Railways, Builders, etc. By ROBERT CHAPMAN. 24mo. $1 50

CALVERT.—LECTURES ON COAL-TAR COLORS AND ON RECENT IMPROVEMENTS AND PROGRESS IN DYEING AND CALICO PRINTING.

Embodying Copious Notes taken at the last London International Exhibition, and *Illustrated with Numerous Patterns of Aniline and other Colors.* By F. GRACE CALVERT, F. R. S., F. C. S. 8vo., cloth $1 50

CRAIK.—THE PRACTICAL AMERICAN MILLWRIGHT AND MILLER.

Comprising the Elementary Principles of Mechanics, Mechanism, and Motive Power, Hydraulics and Hydraulic Motors, Mill-dams, Saw Mills, Grist Mills, the Oat Meal Mill, the Barley Mill, Wool Carding, and Cloth Fulling and Dressing, Wind Mills, Steam Power, &c. By DAVID CRAIK, Millwright. Illustrated by numerous wood engravings, and five folding plates. 1 vol. 8vo. $5 00

CAMPIN.—A PRACTICAL TREATISE ON MECHANICAL ENGINEERING:

Comprising Metallurgy, Moulding, Casting, Forging, Tools, Workshop Machinery, Mechanical Manipulation, Manufacture of Steam-engines, etc. etc. With an Appendix on the Analysis of Iron and Iron Ores. By FRANCIS CAMPIN, C. E. To which are added, Observations on the Construction of Steam Boilers, and Remarks upon Furnaces used for Smoke Prevention; with a Chapter on Explosions. By R. Armstrong, C. E., and John Bourne. Rules for Calculating the Change Wheels for Screws on a Turning Lathe, and for a Wheel-cutting Machine. By J. LA NICCA. Management of Steel, including Forging, Hardening, Tempering, Annealing, Shrinking, and Expansion. And the Case-hardening of Iron. By G. EDE. 8vo. Illustrated with 29 plates and 100 wood engravings.

$6 00

CAMPIN.—THE PRACTICE OF HAND-TURNING IN WOOD, IVORY, SHELL, ETC.:

With Instructions for Turning such works in Metal as may be required in the Practice of Turning Wood, Ivory, etc. Also an Appendix on Ornamental Turning. By FRANCIS CAMPIN, with Numerous Illustrations, 12mo., cloth . . $3 00

CAPRON DE DOLE.—DUSSAUCE.—BLUES AND CARMINES OF INDIGO.

A Practical Treatise on the Fabrication of every Commercial Product derived from Indigo. By FELICIEN CAPRON DE DOLE Translated, with important additions, by Professor H. DUSSAUCE. 12mo. $2 50

CAREY.—THE WORKS OF HENRY C. CAREY:

CONTRACTION OR EXPANSION? REPUDIATION OR RESUMPTION? Letters to Hon. Hugh McCulloch. 8vo. 38

FINANCIAL CRISES, their Causes and Effects. 8vo. paper 25

HARMONY OF INTERESTS; Agricultural, Manufacturing, and Commercial. 8vo., paper $1 00
Do. do. cloth . . . $1 50

LETTERS TO THE PRESIDENT OF THE UNITED STATES. Paper $1 00

MANUAL OF SOCIAL SCIENCE. Condensed from Carey's "Principles of Social Science." By KATE MCKEAN. 1 vol. 12mo. $2 25

MISCELLANEOUS WORKS: comprising "Harmony of Interests," "Money," "Letters to the President," "French and American Tariffs," "Financial Crises," "The Way to Outdo England without Fighting Her," "Resources of the Union," "The Public Debt," "Contraction or Expansion," "Review of the Decade 1857—'67," "Reconstruction," etc. etc. 1 vol. 8vo., cloth $4 50

MONEY: A LECTURE before the N. Y. Geographical and Statistical Society. 8vo., paper 25

PAST, PRESENT, AND FUTURE. 8vo. $2 50

PRINCIPLES OF SOCIAL SCIENCE. 3 volumes 8vo., cloth $10 00

REVIEW OF THE DECADE 1857—'67. 8vo., paper 50

RECONSTRUCTION: INDUSTRIAL, FINANCIAL, AND POLITICAL. Letters to the Hon. Henry Wilson, U. S. S. 8vo paper 50

THE PUBLIC DEBT, LOCAL AND NATIONAL. How to provide for its discharge while lessening the burden of Taxation. Letter to David A. Wells, Esq., U. S. Revenue Commission. 8vo., paper 25

THE RESOURCES OF THE UNION. A Lecture read, Dec. 1865, before the American Geographical and Statistical Society, N. Y., and before the American Association for the Advancement of Social Science, Boston 50

THE SLAVE TRADE, DOMESTIC AND FOREIGN; Why it Exists, and How it may be Extinguished. 12mo., cloth $1 50

THE WAY TO OUTDO ENGLAND WITHOUT FIGHTING HER. Letters to the Hon. Schuyler Colfax. 8vo., paper $1 00

CAMUS.—A TREATISE ON THE TEETH OF WHEELS:
Demonstrating the best forms which can be given to them for the purposes of Machinery, such as Mill-work and Clock-work. Translated from the French of M. CAMUS. By JOHN I. HAWKINS. Illustrated by 40 plates. 8vo. $3 00

CLOUGH.—THE CONTRACTOR'S MANUAL AND BUILDER'S PRICE-BOOK:
Designed to elucidate the method of ascertaining, correctly, the value and Quantity of every description of Work and Materials used in the Art of Building, from their Prime Cost in any part of the United States, collected from extensive experience and observation in Building and Designing; to which are added a large variety of Tables, Memoranda, etc., indispensable to all engaged or concerned in erecting buildings of any kind. By A. B. CLOUGH, Architect, 24mo., cloth 75

COLBURN.—THE GAS-WORKS OF LONDON:
Comprising a sketch of the Gas-works of the city, Process of Manufacture, Quantity Produced, Cost, Profit, etc. By ZERAH COLBURN. 8vo., cloth 75

COLBURN.—THE LOCOMOTIVE ENGINE:
Including a Description of its Structure, Rules for Estimating its Capabilities, and Practical Observations on its Construction and Management. By ZERAH COLBURN. Illustrated. A new edition. 12mo. $1 25

COLBURN AND MAW.—THE WATER-WORKS OF LONDON:
Together with a Series of Articles on various other Waterworks. By ZERAH COLBURN and W. MAW. Reprinted from "Engineering." In one volume, 8vo. . . $4 00

DAGUERREOTYPIST AND PHOTOGRAPHER'S COMPANION:
12mo., cloth $1 25

DUPLAIS.—A COMPLETE TREATISE ON THE DISTILLATION AND PREPARATION OF ALCOHOLIC AND OTHER LIQUORS:
From the French of M. DUPLAIS. Translated and Edited by M. McKENNIE, M. D. Illustrated. 8vo. (*In press.*)

DIRCKS.—PERPETUAL MOTION:

Or Search for Self-Motive Power during the 17th, 18th, and 19th centuries. Illustrated from various authentic sources in Papers, Essays, Letters, Paragraphs, and numerous Patent Specifications, with an Introductory Essay by HENRY DIRCKS, C. E. Illustrated by numerous engravings of machines. 12mo., cloth $3 50

DIXON.—THE PRACTICAL MILLWRIGHT'S AND ENGINEER'S GUIDE:

Or Tables for Finding the Diameter and Power of Cogwheels; Diameter, Weight, and Power of Shafts; Diameter and Strength of Bolts, etc. etc. By THOMAS DIXON. 12mo., cloth. $1 50

DUNCAN.—PRACTICAL SURVEYOR'S GUIDE:

Containing the necessary information to make any person, of common capacity, a finished land surveyor without the aid of a teacher. By ANDREW DUNCAN. Illustrated. 12mo., cloth. $1 25

DUSSAUCE.—A NEW AND COMPLETE TREATISE ON THE ARTS OF TANNING, CURRYING, AND LEATHER DRESSING:

Comprising all the Discoveries and Improvements made in France, Great Britain, and the United States. Edited from Notes and Documents of Messrs. Sallerou, Grouvelle, Duval, Dessables, Labarraque, Payen, René, De Fontenelle, Malapeyre, etc. etc. By Prof H. DUSSAUCE, Chemist. Illustrated by 212 wood engravings. 8vo. $10 00

DUSSAUCE.—A GENERAL TREATISE ON THE MANUFACTURE OF SOAP, THEORETICAL AND PRACTICAL:

Comprising the Chemistry of the Art, a Description of all the Raw Materials and their Uses. Directions for the Establishment of a Soap Factory, with the necessary Apparatus, Instructions in the Manufacture of every variety of Soap, the Assay and Determination of the Value of Alkalies, Fatty Substances, Soaps, etc. etc. By PROFESSOR H. DUSSAUCE. With an Appendix, containing Extracts from the Reports of the International Jury on Soaps, as exhibited in the Paris Universal Exposition, 1867, numerous Tables, etc. etc. Illustrated by engravings. In one volume 8vo. of over 800 pages $10 00

DUSSAUCE.—A PRACTICAL GUIDE FOR THE PERFUMER:
Being a New Treatise on Perfumery the most favorable to the Beauty without being injurious to the Health, comprising a Description of the substances used in Perfumery, the Formulæ of more than one thousand Preparations, such as Cosmetics, Perfumed Oils, Tooth Powders, Waters, Extracts, Tinctures, Infusions, Vinaigres, Essential Oils, Pastels, Creams, Soaps, and many new Hygienic Products not hitherto described. Edited from Notes and Documents of Messrs. Debay, Lunel, etc. With additions by Professor H. Dussauce, Chemist. 12mo. $3 00

DUSSAUCE.—PRACTICAL TREATISE ON THE FABRICATION OF MATCHES, GUN COTTON, AND FULMINATING POWDERS.
By Professor H. Dussauce. 12mo. $3 00

DUSSAUCE.—A GENERAL TREATISE ON THE MANUFACTURE OF VINEGAR, THEORETICAL AND PRACTICAL.
Comprising the various methods, by the slow and the quick processes, with Alcohol, Wine, Grain, Cider, and Molasses, as well as the Fabrication of Wood Vinegar, etc. By Prof. H. Dussauce. 12mo. (*In press.*)

DE GRAFF.—THE GEOMETRICAL STAIR-BUILDERS' GUIDE:
Being a Plain Practical System of Hand-Railing, embracing all its necessary Details, and Geometrically Illustrated by 22 Steel Engravings; together with the use of the most approved principles of Practical Geometry. By Simon De Graff, Architect. 4to. $5 00

DYER AND COLOR-MAKER'S COMPANION:
Containing upwards of two hundred Receipts for making Colors, on the most approved principles, for all the various styles and fabrics now in existence; with the Scouring Process, and plain Directions for Preparing, Washing-off, and Finishing the Goods. In one vol. 12mo. $1 25

EASTON.—A PRACTICAL TREATISE ON STREET OR HORSE-POWER RAILWAYS:
Their Location, Construction, and Management; with General Plans and Rules for their Organization and Operation; together with Examinations as to their Comparative Advantages over the Omnibus System, and Inquiries as to their Value for Investment; including Copies of Municipal Ordinances relating thereto. By Alexander Easton, C. E. Illustrated by 23 plates, 8vo., cloth $2 00

FORSYTH.—BOOK OF DESIGNS FOR HEAD-STONES, MURAL, AND OTHER MONUMENTS:

Containing 78 Elaborate and Exquisite Designs. By FORSYTH. 4to. (*In press.*)

FAIRBAIRN.—THE PRINCIPLES OF MECHANISM AND MACHINERY OF TRANSMISSION:

Comprising the Principles of Mechanism, Wheels, and Pulleys, Strength and Proportions of Shafts, Couplings of Shafts, and Engaging and Disengaging Gear. By WILLIAM FAIRBAIRN, Esq., C. E., LL. D., F. R. S., F. G. S., Corresponding Member of the National Institute of France, and of the Royal Academy of Turin; Chevalier of the Legion of Honor, etc. etc. Beautifully illustrated by over 150 wood-cuts. In one volume 12mo. $2 50

FAIRBAIRN.—PRIME-MOVERS:

Comprising the Accumulation of Water-power; the Construction of Water-wheels and Turbines; the Properties of Steam; the Varieties of Steam-engines and Boilers and Wind-mills. By WILLIAM FAIRBAIRN, C. E., LL. D., F. R. S., F. G. S. Author of "Principles of Mechanism and the Machinery of Transmission." With Numerous Illustrations. In one volume. (In press.)

FLAMM.—A PRACTICAL GUIDE TO THE CONSTRUCTION OF ECONOMICAL HEATING APPLICATIONS FOR SOLID AND GASEOUS FUELS:

With the Application of Concentrated Heat, and on Waste Heat, for the Use of Engineers, Architects, Stove and Furnace Makers, Manufacturers of Fire Brick, Zinc, Porcelain, Glass, Earthenware, Steel, Chemical Products, Sugar Refiners, Metallurgists, and all others employing Heat. By M. PIERRE FLAMM, Manufacturer. Illustrated. Translated from the French. One volume, 12mo. (In press.)

GILBART.—A PRACTICAL TREATISE ON BANKING:

By JAMES WILLIAM GILBART. To which is added: THE NATIONAL BANK ACT AS NOW IN FORCE. 8vo. . . $4 50

GESNER.—A PRACTICAL TREATISE ON COAL, PETROLEUM, AND OTHER DISTILLED OILS.

By ABRAHAM GESNER, M. D., F. G. S. Second edition, revised and enlarged. By GEORGE WELTDEN GESNER, Consulting Chemist and Engineer. Illustrated. 8vo. . . $3 50

GOTHIC ALBUM FOR CABINET MAKERS:

Comprising a Collection of Designs for Gothic Furniture. Illustrated by twenty-three large and beautifully engraved plates. Oblong $3 00

GRANT.—BEET-ROOT SUGAR AND CULTIVATION OF THE BEET:

By E. B. GRANT. 12mo. $1 25

GREGORY.—MATHEMATICS FOR PRACTICAL MEN:

Adapted to the Pursuits of Surveyors, Architects, Mechanics, and Civil Engineers. By OLINTHUS GREGORY. 8vo., plates, cloth $3 00

GRISWOLD.—RAILROAD ENGINEER'S POCKET COMPANION.

Comprising Rules for Calculating Deflection Distances and Angles, Tangential Distances and Angles, and all Necessary Tables for Engineers; also the art of Levelling from Preliminary Survey to the Construction of Railroads, intended Expressly for the Young Engineer, together with Numerous Valuable Rules and Examples. By W. GRISWOLD. 12mo., tucks. $1 75

GUETTIER.—METALLIC ALLOYS:

Being a Practical Guide to their Chemical and Physical Properties, their Preparation, Composition, and Uses. Translated from the French of A. GUETTIER, Engineer and Director of Founderies, author of "La Fouderie en France," etc. etc. By A. A. FESQUET, Chemist and Engineer. In one volume, 12mo. (In press, *shortly to be published.*)

HATS AND FELTING:

A Practical Treatise on their Manufacture. By a Practical Hatter. Illustrated by Drawings of Machinery, &c., 8vo. $1 25

HAY.—THE INTERIOR DECORATOR:

The Laws of Harmonious Coloring adapted to Interior Decorations: with a Practical Treatise on House-Painting. By D. R. HAY, House-Painter and Decorator. Illustrated by a Diagram of the Primary, Secondary, and Tertiary Colors. 12mo. $2 25

HUGHES.—AMERICAN MILLER AND MILLWRIGHT'S ASSISTANT:

By WM. CARTER HUGHES. A new edition. In one volume, 12mo. $1 50

HUNT.—THE PRACTICE OF PHOTOGRAPHY.

By Robert Hunt, Vice-President of the Photographic Society, London, with numerous illustrations. 12mo., cloth . 75

HURST.—A HAND-BOOK FOR ARCHITECTURAL SURVEYORS:

Comprising Formulæ useful in Designing Builder's work, Table of Weights, of the materials used in Building, Memoranda connected with Builders' work, Mensuration, the Practice of Builders' Measurement, Contracts of Labor, Valuation of Property, Summary of the Practice in Dilapidation, etc. etc. By J. F. Hurst, C. E. 2d edition, pocket-book form, full bound $2 50

JERVIS.—RAILWAY PROPERTY:

A Treatise on the Construction and Management of Railways; designed to afford useful knowledge, in the popular style, to the holders of this class of property; as well as Railway Managers, Officers, and Agents. By John B. Jervis, late Chief Engineer of the Hudson River Railroad, Croton Aqueduct, &c. One vol. 12mo., cloth $2 00

JOHNSON.—A REPORT TO THE NAVY DEPARTMENT OF THE UNITED STATES ON AMERICAN COALS:

Applicable to Steam Navigation and to other purposes. By Walter R. Johnson. With numerous illustrations. 607 pp. 8vo., half morocco . . . .

JOHNSON.—THE COAL TRADE OF BRITISH AMERICA:

With Researches on the Characters and Practical Values of American and Foreign Coals. By Walter R. Johnson, Civil and Mining Engineer and Chemist. 8vo. . . . .

JOHNSTON.—INSTRUCTIONS FOR THE ANALYSIS OF SOILS, LIMESTONES, AND MANURES.

By J. W. F. Johnston. 12mo. 38

KEENE.—A HAND-BOOK OF PRACTICAL GAUGING,

For the Use of Beginners, to which is added A Chapter on Distillation, describing the process in operation at the Custom House for ascertaining the strength of wines. By James B. Keene, of H. M. Customs. 8vo. $1 25

KENTISH.—A TREATISE ON A BOX OF INSTRUMENTS,

And the Slide Rule; with the Theory of Trigonometry and Logarithms, including Practical Geometry, Surveying, Measuring of Timber, Cask and Malt Gauging, Heights, and Distances. By Thomas Kentish. In one volume. 12mo. . $1 25

KOBELL.—ERNI.—MINERALOGY SIMPLIFIED:

A short method of Determining and Classifying Minerals, by means of simple Chemical Experiments in the Wet Way. Translated from the last German Edition of F. VON KOBELL, with an Introduction to Blowpipe Analysis and other additions. By HENRI ERNI, M. D., Chief Chemist, Department of Agriculture, author of "Coal Oil and Petroleum." In one volume, 12mo. $2 50

LAFFINEUR.—A PRACTICAL GUIDE TO HYDRAULICS FOR TOWN AND COUNTRY;

Or a Complete Treatise on the Building of Conduits for Water for Cities, Towns, Farms, Country Residences, Workshops, etc. Comprising the means necessary for obtaining at all times abundant supplies of Drinkable Water. Translated from the French of M. JULES LAFFINEUR, C. E. Illustrated. (In press.)

LANDRIN.—A TREATISE ON STEEL:

Comprising its Theory, Metallurgy, Properties, Practical Working, and Use. By M. H. C. LANDRIN, Jr., Civil Engineer. Translated from the French, with Notes, by A. A. FESQUET, Chemist and Engineer. With an Appendix on the Bessemer and the Martin Processes for Manufacturing Steel, from the Report of Abram S. Hewitt, United States Commissioner to the Universal Exposition, Paris, 1867. 12mo. $3 00

LARKIN.—THE PRACTICAL BRASS AND IRON FOUNDER'S GUIDE:

A Concise Treatise on Brass Founding, Moulding, the Metals and their Alloys, etc.; to which are added Recent Improvements in the Manufacture of Iron, Steel by the Bessemer Process, etc. etc. By JAMES LARKIN, late Conductor of the Brass Foundry Department in Reany, Neafie & Co.'s Penn Works, Philadelphia. Fifth edition, revised, with Extensive additions. In one volume, 12mo. $2 25

LEAVITT.—FACTS ABOUT PEAT AS AN ARTICLE OF FUEL:
With Remarks upon its Origin and Composition, the Localities in which it is found, the Methods of Preparation and Manufacture, and the various Uses to which it is applicable; together with many other matters of Practical and Scientific Interest. To which is added a chapter on the Utilization of Coal Dust with Peat for the Production of an Excellent Fuel at Moderate Cost, especially adapted for Steam Service. By H. T. LEAVITT. Third edition. 12mo. $1 75

LEROUX.—A PRACTICAL TREATISE ON THE MANUFACTURE OF WORSTEDS AND CARDED YARNS:
Translated from the French of CHARLES LEROUX, Mechanical Engineer, and Superintendent of a Spinning Mill. By Dr. H. PAINE, and A. A. FESQUET. Illustrated by 12 large plates. In one volume 8vo. $5 00

LESLIE (MISS).—COMPLETE COOKERY:
Directions for Cookery in its Various Branches. By MISS LESLIE. 60th edition. Thoroughly revised, with the addition of New Receipts. In 1 vol. 12mo., cloth . . $1 50

LESLIE (MISS). LADIES' HOUSE BOOK:
a Manual of Domestic Economy. 20th revised edition. 12mo., cloth $1 25

LESLIE (MISS).—TWO HUNDRED RECEIPTS IN FRENCH COOKERY.
12mo. 50

LIEBER.—ASSAYER'S GUIDE:
Or, Practical Directions to Assayers, Miners, and Smelters, for the Tests and Assays, by Heat and by Wet Processes, for the Ores of all the principal Metals, of Gold and Silver Coins and Alloys, and of Coal, etc. By OSCAR M. LIEBER. 12mo., cloth $1 25

LOVE.—THE ART OF DYEING, CLEANING, SCOURING, AND FINISHING:
On the most approved English and French methods; being Practical Instructions in Dyeing Silks, Woollens, and Cottons, Feathers, Chips, Straw, etc.; Scouring and Cleaning Bed and Window Curtains, Carpets, Rugs, etc.; French and English Cleaning, etc. By THOMAS LOVE. Second American Edition, to which are added General Instructions for the Use of Aniline Colors. 8vo. 5 00

MAIN AND BROWN.—QUESTIONS ON SUBJECTS CONNECTED WITH THE MARINE STEAM-ENGINE:

And Examination Papers; with Hints for their Solution. By THOMAS J. MAIN, Professor of Mathematics, Royal Naval College, and THOMAS BROWN, Chief Engineer, R. N. 12mo., cloth $1 50

MAIN AND BROWN.—THE INDICATOR AND DYNAMOMETER:

With their Practical Applications to the Steam-Engine. By THOMAS J. MAIN, M. A. F. R., Ass't Prof. Royal Naval College, Portsmouth, and THOMAS BROWN, Assoc. Inst. C. E., Chief Engineer, R. N., attached to the R. N. College. Illustrated. From the Fourth London Edition. 8vo. $1 50

MAIN AND BROWN.—THE MARINE STEAM-ENGINE.

By THOMAS J. MAIN, F. R. Ass't S. Mathematical Professor at Royal Naval College, and THOMAS BROWN, Assoc. Inst. C. E. Chief Engineer, R. N. Attached to the Royal Naval College. Authors of "Questions Connected with the Marine Steam-Engine," and the "Indicator and Dynamometer." With numerous Illustrations. In one volume 8vo. $5 00

MARTIN.—SCREW-CUTTING TABLES, FOR THE USE OF MECHANICAL ENGINEERS:

Showing the Proper Arrangement of Wheels for Cutting the Threads of Screws of any required Pitch; with a Table for Making the Universal Gas-Pipe Thread and Taps. By W. A. MARTIN, Engineer. 8vo. 50

MILES—A PLAIN TREATISE ON HORSE-SHOEING.

With Illustrations. By WILLIAM MILES, author of "The Horse's Foot" $1 00

MOLESWORTH.—POCKET-BOOK OF USEFUL FORMULÆ AND MEMORANDA FOR CIVIL AND MECHANICAL ENGINEERS.

By GUILFORD L. MOLESWORTH, Member of the Institution of Civil Engineers, Chief Resident Engineer of the Ceylon Railway. Second American from the Tenth London Edition. In one volume, full bound in pocket-book form $2 00

MOORE.—THE INVENTOR'S GUIDE:

Patent Office and Patent Laws: or, a Guide to Inventors, and a Book of Reference for Judges, Lawyers, Magistrates, and others. By J G. MOORE. 12mo., cloth $1 25

NAPIER.—A MANUAL OF ELECTRO-METALLURGY:

Including the Application of the Art to Manufacturing Processes. By JAMES NAPIER. Fourth American, from the Fourth London edition, revised and enlarged. Illustrated by engravings. In one volume, 8vo. $2 00

NAPIER.—A SYSTEM OF CHEMISTRY APPLIED TO DYEING:

By JAMES NAPIER, F. C. S. A New and Thoroughly Revised Edition, completely brought up to the present state of the Science, including the Chemistry of Coal Tar Colors. By A. A. FESQUET, Chemist and Engineer. With an Appendix on Dyeing and Calico Printing, as shown at the Paris Universal Exposition of 1867, from the Reports of the International Jury, etc. Illustrated. In one volume 8vo., 400 pages $5 00

NEWBERY.—GLEANINGS FROM ORNAMENTAL ART OF EVERY STYLE;

Drawn from Examples in the British, South Kensington, Indian, Crystal Palace, and other Museums, the Exhibitions of 1851 and 1862, and the best English and Foreign works. In a series of one hundred exquisitely drawn Plates, containing many hundred examples. By ROBERT NEWBERY. 4to. $15 00

NICHOLSON.—A MANUAL OF THE ART OF BOOK-BINDING:

Containing full instructions in the different Branches of Forwarding, Gilding, and Finishing. Also, the Art of Marbling Book-edges and Paper. By JAMES B. NICHOLSON. Illustrated. 12mo. cloth $2 25

NORRIS.—A HAND-BOOK FOR LOCOMOTIVE ENGINEERS AND MACHINISTS:

Comprising the Proportions and Calculations for Constructing Locomotives; Manner of Setting Valves; Tables of Squares, Cubes, Areas, etc. etc. By SEPTIMUS NORRIS, Civil and Mechanical Engineer. New edition. Illustrated, 12mo., cloth $2 00

NYSTROM.—ON TECHNOLOGICAL EDUCATION AND THE CONSTRUCTION OF SHIPS AND SCREW PROPELLERS:

For Naval and Marine Engineers. By JOHN W. NYSTROM, late Acting Chief Engineer U. S. N. Second edition, revised with additional matter. Illustrated by seven engravings. 12mo. $2 50

O'NEILL.—A DICTIONARY OF DYEING AND CALICO PRINTING:

Containing a brief account of all the Substances and Processes in use in the Art of Dyeing and Printing Textile Fabrics: with Practical Receipts and Scientific Information. By CHARLES O'NEILL, Analytical Chemist; Fellow of the Chemical Society of London; Member of the Literary and Philosophical Society of Manchester; Author of "Chemistry of Calico Printing and Dyeing." To which is added An Essay on Coal Tar Colors and their Application to

Dyeing and Calico Printing. By A. A. FESQUET, Chemist and Engineer. With an Appendix on Dyeing and Calico Printing, as shown at the Exposition of 1867, from the Reports of the International Jury, etc. In one volume 8vo., 491 pages . . $6 00

OSBORN.—THE METALLURGY OF IRON AND STEEL:
Theoretical and Practical: In all its Branches; With Special Reference to American Materials and Processes. By H. S. OSBORN, LL. D., Professor of Mining and Metallurgy in Lafayette College, Easton, Pa. Illustrated by 230 Engravings on Wood, and 6 Folding Plates. 8vo., 972 pages $10 00

OSBORN.—AMERICAN MINES AND MINING:
Theoretically and Practically Considered. By Prof. H. S. OSBORN, Illustrated by numerous engravings. 8vo. (*In preparation.*)

PAINTER, GILDER, AND VARNISHER'S COMPANION:
Containing Rules and Regulations in everything relating to the Arts of Painting, Gilding, Varnishing, and Glass Staining, with numerous useful and valuable Receipts; Tests for the Detection of Adulterations in Oils and Colors, and a statement of the Diseases and Accidents to which Painters, Gilders, and Varnishers are particularly liable, with the simplest methods of Prevention and Remedy. With Directions for Graining, Marbling, Sign Writing, and Gilding on Glass. To which are added COMPLETE INSTRUCTIONS FOR COACH PAINTING AND VARNISHING. 12mo., cloth, $1 50

PALLETT.—THE MILLER'S, MILLWRIGHT'S, AND ENGINEER'S GUIDE.
By HENRY PALLETT. Illustrated. In one vol. 12mo. . $3 00

PERKINS.—GAS AND VENTILATION.
Practical Treatise on Gas and Ventilation. With Special Relation to Illuminating, Heating, and Cooking by Gas. Including Scientific Helps to Engineer-students and others. With illustrated Diagrams. By E. E. PERKINS. 12mo., cloth . . . $1 25

PERKINS AND STOWE.—A NEW GUIDE TO THE SHEET-IRON AND BOILER PLATE ROLLER:
Containing a Series of Tables showing the Weight of Slabs and Piles to Produce Boiler Plates, and of the Weight of Piles and the Sizes of Bars to Produce Sheet-iron; the Thickness of the Bar Gauge in Decimals; the Weight per foot, and the Thickness on the Bar or Wire Gauge of the fractional parts of an inch; the Weight per sheet, and the Thickness on the Wire Gauge of Sheet-iron of various dimensions to weigh 112 lbs. per bundle; and the conversion of Short Weight into Long Weight, and Long Weight into Short. Estimated and collected by G. H. PERKINS and J. G. STOWE $2 50

PHILLIPS AND DARLINGTON.—RECORDS OF MINING AND METALLURGY:

Or, Facts and Memoranda for the use of the Mine Agent and Smelter. By J. ARTHUR PHILLIPS, Mining Engineer, Graduate of the Imperial School of Mines, France, etc., and JOHN DARLINGTON. Illustrated by numerous engravings. In one vol. 12mo. . $2 00

PRADAL, MALEPEYRE, AND DUSSAUCE.—A COMPLETE TREATISE ON PERFUMERY:

Containing notices of the Raw Material used in the Art, and the Best Formulæ. According to the most approved Methods followed in France, England, and the United States. By M. P. PRADAL, Perfumer-Chemist, and M. F. MALEPEYRE. Translated from the French, with extensive additions, by Prof. H. DUSSAUCE. 8vo. $10

PROTEAUX.—PRACTICAL GUIDE FOR THE MANUFACTURE OF PAPER AND BOARDS.

By A. PROTEAUX, Civil Engineer, and Graduate of the School of Arts and Manufactures, Director of Thiers's Paper Mill, 'Puy-de-Dômé. With additions, by L. S. LE NORMAND. Translated from the French, with Notes, by HORATIO PAINE, A. B., M. D. To which is added a Chapter on the Manufacture of Paper from Wood in the United States, by HENRY T. BROWN, of the "American Artisan." Illustrated by six plates, containing Drawings of Raw Materials, Machinery, Plans of Paper-Mills, etc. etc. 8vo. $5 00

REGNAULT.—ELEMENTS OF CHEMISTRY.

By M. V. REGNAULT. Translated from the French by T. FORREST BENTON, M. D., and edited, with notes, by JAMES C. BOOTH, Melter and Refiner U. S. Mint, and WM. L. FABER, Metallurgist and Mining Engineer. Illustrated by nearly 700 wood engravings. Comprising nearly 1500 pages. In two vols. 8vo., cloth $10 00

REID.—A PRACTICAL TREATISE ON THE MANUFACTURE OF PORTLAND CEMENT:

By HENRY REID, C. E. To which is added a Translation of M. A. Lipowitz's Work, describing a new method adopted in Germany of Manufacturing that Cement. By W. F. REID. Illustrated by plates and wood engravings. 8vo. $7 00

RIFFAULT, VERGNAUD, AND TOUSSAINT.—A PRACTICAL TREATISE ON THE MANUFACTURE OF COLORS FOR PAINTING:

Containing the best Formulæ and the Processes the Newest and in most General Use. By MM. RIFFAULT, VERGNAUD, and TOUSSAINT. Revised and Edited by M. F. MALEPEYRE and Dr. EMIL WINCKLER. Illustrated by Engravings. In one vol. 8vo. (*In preparation.*)

RIFFAULT, VERGNAUD, AND TOUSSAINT.—A PRACTICAL TREATISE ON THE MANUFACTURE OF VARNISHES:

By MM. RIFFAULT, VERGNAUD, and TOUSSAINT. Revised and Edited by M. F. MALEPEYRE and Dr. EMIL WINCKLER. Illustrated. In one vol. 8vo. (*In preparation.*)

SHUNK.—A PRACTICAL TREATISE ON RAILWAY CURVES AND LOCATION, FOR YOUNG ENGINEERS.

By WM. F. SHUNK, Civil Engineer. 12mo., tucks . . $2 00

SMEATON.—BUILDER'S POCKET COMPANION:

Containing the Elements of Building, Surveying, and Architecture; with Practical Rules and Instructions connected with the subject. By A. C. SMEATON, Civil Engineer, etc. In one volume, 12mo. $1 50

SMITH.—THE DYER'S INSTRUCTOR:

Comprising Practical Instructions in the Art of Dyeing Silk, Cotton, Wool, and Worsted, and Woollen Goods: containing nearly 800 Receipts. To which is added a Treatise on the Art of Padding; and the Printing of Silk Warps, Skeins, and Handkerchiefs, and the various Mordants and Colors for the different styles of such work. By DAVID SMITH, Pattern Dyer, 12mo., cloth $3 00

SMITH.—THE PRACTICAL DYER'S GUIDE:

Comprising Practical Instructions in the Dyeing of Shot Cobourgs, Silk Striped Orleans, Colored Orleans from Black Warps, ditto from White Warps, Colored Cobourgs from White Warps, Merinos, Yarns, Woollen Cloths, etc. Containing nearly 300 Receipts, to most of which a Dyed Pattern is annexed. Also, a Treatise on the Art of Padding. By DAVID SMITH. In one vol. 8vo. $25 00

SHAW.—CIVIL ARCHITECTURE:

Being a Complete Theoretical and Practical System of Building, containing the Fundamental Principles of the Art. By EDWARD SHAW, Architect. To which is added a Treatise on Gothic Architecture, &c. By THOMAS W. SILLOWAY and GEORGE M. HARDING, Architects. The whole illustrated by 102 quarto plates finely engraved on copper. Eleventh Edition. 4to. Cloth. $10 00

SLOAN.—AMERICAN HOUSES:

A variety of Original Designs for Rural Buildings. Illustrated by 26 colored Engravings, with Descriptive References. By SAMUEL SLOAN, Architect, author of the "Model Architect," etc. etc. 8vo. $2 50

SMITH.—PARKS AND PLEASURE GROUNDS:

Or, Practical Notes on Country Residences, Villas, Public Parks, and Gardens. By CHARLES H. J. SMITH, Landscape Gardener and Garden Architect, etc. etc. 12mo. $2 25

STOKES.—CABINET-MAKER'S AND UPHOLSTERER'S COMPANION:

Comprising the Rudiments and Principles of Cabinet-making and Upholstery, with Familiar Instructions, Illustrated by Examples for attaining a Proficiency in the Art of Drawing, as applicable to Cabinet-work; The Processes of Veneering, Inlaying, and Buhl-work; the Art of Dyeing and Staining Wood, Bone, Tortoise Shell, etc. Directions for Lackering, Japanning, and Varnishing; to make French Polish; to prepare the Best Glues, Cements, and Compositions, and a number of Receipts, particularly for workmen generally. By J. STOKES. In one vol. 12mo. With illustrations $1 25

STRENGTH AND OTHER PROPERTIES OF METALS.

Reports of Experiments on the Strength and other Properties of Metals for Cannon. With a Description of the Machines for Testing Metals, and of the Classification of Cannon in service. By Officers of the Ordnance Department U. S. Army. By authority of the Secretary of War. Illustrated by 25 large steel plates. In 1 vol. quarto $10 00

TABLES SHOWING THE WEIGHT OF ROUND, SQUARE, AND FLAT BAR IRON, STEEL, ETC.

By Measurement. Cloth 63

TAYLOR.—STATISTICS OF COAL:

Including Mineral Bituminous Substances employed in Arts and Manufactures; with their Geographical, Geological, and Commercial Distribution and amount of Production and Consumption on the American Continent. With Incidental Statistics of the Iron Manufacture. By R. C. TAYLOR. Second edition, revised by S. S. HALDEMAN. Illustrated by five Maps and many wood engravings. 8vo., cloth $6 00

TEMPLETON.—THE PRACTICAL EXAMINATOR ON STEAM AND THE STEAM-ENGINE:

With Instructive References relative thereto, for the Use of Engineers, Students, and others. By WM. TEMPLETON, Engineer 12mo. $1 25

THOMAS.—THE MODERN PRACTICE OF PHOTOGRAPHY.

By R. W. THOMAS, F. C. S. 8vo., cloth 75

THOMSON.—FREIGHT CHARGES CALCULATOR.
By ANDREW THOMSON, Freight Agent $1 25

TURNING: SPECIMENS OF FANCY TURNING EXECUTED ON THE HAND OR FOOT LATHE:
With Geometric, Oval, and Eccentric Chucks, and Elliptical Cutting Frame. By an Amateur. Illustrated by 30 exquisite Photographs. 4to. $3 00

TURNER'S (THE) COMPANION:
Containing Instructions in Concentric, Elliptic, and Eccentric Turning; also various Plates of Chucks, Tools, and Instruments; and Directions for using the Eccentric Cutter, Drill, Vertical Cutter, and Circular Rest; with Patterns and Instructions for working them. A new edition in 1 vol. 12mo. $1 50

URBIN—BRULL.—A PRACTICAL GUIDE FOR PUDDLING IRON AND STEEL.
By ED. URBIN, Engineer of Arts and Manufactures. A Prize Essay read before the Association of Engineers, Graduate of the School of Mines, of Liege, Belgium, at the Meeting of 1865–6. To which is added a COMPARISON OF THE RESISTING PROPERTIES OF IRON AND STEEL. By A. BRULL. Translated from the French by A. A. FESQUET, Chemist and Engineer. In one volume, 8vo. $1 00

WARN.—THE SHEET METAL WORKER'S INSTRUCTOR, FOR ZINC, SHEET-IRON, COPPER AND TIN PLATE WORKERS, &c.
By REUBEN HENRY WARN, Practical Tin Plate Worker. Illustrated by 32 plates and 37 wood engravings. 8vo. . . $3 00

WATSON.—A MANUAL OF THE HAND-LATHE.
By EGBERT P. WATSON, Late of the "Scientific American," Author of "Modern Practice of American Machinists and Engineers," In one volume, 12mo. $1 50

WATSON.—THE MODERN PRACTICE OF AMERICAN MACHINISTS AND ENGINEERS:
Including the Construction, Application, and Use of Drills, Lathe Tools, Cutters for Boring Cylinders, and Hollow Work Generally, with the most Economical Speed of the same, the Results verified by Actual Practice at the Lathe, the Vice, and on the Floor. Together with Workshop management, Economy of Manufacture, the Steam-Engine, Boilers, Gears, Belting, etc. etc. By EGBERT P. WATSON, late of the "Scientific American." Illustrated by eighty-six engravings. 12mo. $2 50

WATSON.—THE THEORY AND PRACTICE OF THE ART OF WEAVING BY HAND AND POWER:

With Calculations and Tables for the use of those connected with the Trade. By JOHN WATSON, Manufacturer and Practical Machine Maker. Illustrated by large drawings of the best Power-Looms. 8vo. $10 00

WEATHERLY.—TREATISE ON THE ART OF BOILING SUGAR, CRYSTALLIZING, LOZENGE-MAKING, COMFITS, GUM GOODS,

And other processes for Confectionery, &c. In which are explained, in an easy and familiar manner, the various Methods of Manufacturing every description of Raw and Refined Sugar Goods, as sold by Confectioners and others $2 00

WILL.—TABLES FOR QUALITATIVE CHEMICAL ANALYSIS.

By Prof. HEINRICH WILL, of Giessen, Germany. Seventh edition. Translated by CHARLES F. HIMES, Ph. D., Professor of Natural Science, Dickinson College, Carlisle, Pa. . . $1 25

WILLIAMS.—ON HEAT AND STEAM:

Embracing New Views of Vaporization, Condensation, and Expansion. By CHARLES WYE WILLIAMS, A. I. C. E. Illustrated. 8vo. $3 50

WOHLER.—A PRACTICAL TREATISE ON ANALYTICAL CHEMISTRY.

By F. Wöhler. With additions by GRANDEAU and TROOST. Edited by H. B. NASON, Professor of Chemistry, Rensselaer Institute, Troy, N. Y. With numerous Illustrations. (*In press.*)

WORSSAM.—ON MECHANICAL SAWS:

From the Transactions of the Society of Engineers, 1867. By S. W. WORSSAM, Jr. Illustrated by 18 large folding plates. 8vo. $5 00

ARLOT.—A COMPLETE GUIDE FOR COACH PAINTERS.

Translated from the French of M. ARLOT, Coach Painter; late Master Painter for eleven years with M. Ehrler, Coach Manufacturer, Paris. With important American additions. (*In press.*)

VOGDES.—THE ARCHITECT'S AND BUILDER'S POCKET COMPANION.

By F. W. VOGDES, Architect. Illustrated. Full bound in pocket-book form. (*In press.*)

www.ingramcontent.com/pod-product-compliance
Lightning Source LLC
LaVergne TN
LVHW021130110826
845150LV00005B/978

* 9 7 8 1 4 2 5 5 5 0 7 2 1 *